# INTEGRATIVE NEUROSCIENCE

**To Adam and Kerry and the children of all brain scientists.**

Jim Wright and Walter Freeman, who have pointed the way forward for modeling and understanding the brain as a dynamical system. Chris Rennie, who exemplifies the future "Integrative Neuroscientist" — a physicist who has embraced neuroscience and is able to model details and concepts about the brain in a bold yet step by step manner.

# INTEGRATIVE NEUROSCIENCE

## BRINGING TOGETHER BIOLOGICAL, PSYCHOLOGICAL AND CLINICAL MODELS OF THE HUMAN BRAIN

Edited by Evian Gordon
Senior Lecturer, Department of Psychological Medicine, University of Sydney
and
Director, Brain Dynamics Centre, Westmead Hospital, Australia

**harwood academic publishers**
Australia • Canada • France • Germany • India • Japan
Luxembourg • Malaysia • The Netherlands • Russia • Singapore
Switzerland

Amsteldijk 166
1st Floor
1079 LH Amsterdam
The Netherlands

---

**British Library Cataloguing in Publication Data**

A catalogue record for this book is available from the British Library.

ISBN 90-5823-055-4 soft cover

# TABLE OF CONTENTS

# LIST OF CONTRIBUTORS

Dr Ken Ashwell
School of Anatomy
The University of NSW
Kensington NSW 2052
Australia

Professor Phil Beart
Department of Pharmacology
Monash University
Clayton, Victoria 3168
Australia

Dr Alan Freeman
Department of Biomedical Sciences
The University of Sydney
Cumberland Campus
Lidcombe  NSW 2141
Australia

Professor Walter Freeman
Department of MCB-Neu LSA 129
University of California
Berkeley CA 94720-3200
USA

Dr Victor Fung
Department of Neurology
Westmead Hospital
Darcy Road
Westmead NSW 2145
Australia

Dr Glenn Hunt
Concord Hospital
Department of Psychological Medicine
Clinical Sciences Building
Concord  NSW  2139
Australia

Dr Ian McGregor
Department of Psychology
A19, University of Sydney
Sydney NSW 2006
Australia

Professor John Mazziotta
Head, Department of Neurology
UCLA School of Medicine
Reed Neurological Research
710 Westwood Plz
Los Angeles, CA 90024-1764
USA

Professor Russell Meares
Academic Head
Professor of Psychiatry
Westmead Hospital
Westmead NSW 2145
Australia

Professor John Morris
Department of Neurology
Westmead Hospital
Darcy Road
Westmead NSW 2145
Australia

Professor George Paxinos
School of Anatomy
The University of  New South Wales
Kensington  NSW  2052
Australia

Mr Chris Rennie
Department of Medical physics
Westmead Hospital
Westmead  NSW 2145
Australia

Dr William Schmidt
Department of Psychology
Dalhousie University
Halifax, Nova Scotia, B3H4J1
Canada

Dr Peter Slezak
School of Science and Technology Studies
University of NSW
Kensington NSW 2052
Australia

Professor Graeme Smith
Monash Medical Centre
Department of Pscyhological Medicine
246 Clayton Road
Victoria 3168 Australia

Dr Elizabeth Tancred
School of Anatomy
University of New South Wales
Kensington NSW 2052
Australia

Professor Phillip Tobias
Department of Anatomical Sciences
University of the Witwatersrand
7 York Road
Parktown 2193
South Africa

Dr Arthur Toga
Laboratory of Neuro Imaging
UCLA School of Medicine
Reed Neurological Research Centre
710 Westwood Plz,
Los Angeles, CA 90024-1769
USA

Professor Jim Wright
Director Brain Dynamics Laboratory
Mental Health Research Institute
Locked Bag 11
Parkville Victoria 3052
Australia

# ACKNOWLEDGEMENTS

To all 25 participants of the 14-part TV series that I produced ("Models of The Human Brain: A multidisciplinary perspective") on which this book is based. Tim Blanche (the associate producer of the TV series); my son Adam for allowing me to use his face to demonstrate the brain models; the brain images generously supplied by Arthur Toga (UCLA Laboratary of Neuroimaging), Wieslaw Nowinski and Brian Scanlan (Thieme Publishers of The Electronic Clinical Brain Atlas, Multiplanar Navigation of the Human Brain), Karl Heinz Hoehne (Voxel-Man, Brain and Skull, Springer Electronic Media); Steen Rees for his skills and multimedia network.

My multidisciplinary collaborators (including a number of the authors of chapters) who helped shape the ideas or the execution of this book in ways too numerous to detail: Homayoun Bahramali, Michael Breakspear, Ian Baguley, Robert Barry, Kerri Brown, Richard Bryant, Simon Clarke, Edyta Delfel, Freeda Diab, Kim Felmingham, Sascha Frydman, Hazel Grant, Craig Gonsalvez, Albert Haig, Anthony Harris, Kaye Horley, Anne Kamal, Carmen Kendall, Michael Kohn, Kris Kozek, Sean Kristoffersen, Jim Lagopoulos, Wai Man Li, Ilario Lazzaro, George Lianos, Lee Lim, Russell Meares, John Morris, Barry Manor, Dianne Marshall, Dimitri Melkonyan, Lena Melville, Peter Robinson, Glynn Rogers, Jody Tröstel, Heather White, Stephanie Whitmont and Lea Williams.

**Chapter 1**

# INTEGRATIVE NEUROSCIENCE: THE BIG PICTURE

Evian Gordon

The Brain Dynamics Centre, Westmead Hospital and
Department of Psychological Medicine, The University of Sydney

The human brain's 100 billion neurons and trillions of connections serve to vigilantly keep us out of danger in a changing environment, provide us with the skills to satisfy our material needs, operate within a social system, soothe our emotional needs and contemplate our sense of self.

These processes have been shaped by evolutionary adaptive forces of natural selection and modulated by culture and experience. Their complexity is daunting. A combinatorial explosion of interacting functional mechanisms (for example, time courses of synaptic activity that span milliseconds to years) resides in each and every neuron. From these microscopic mechanisms emerge neural networks and whole brain phenomena of increasing order and purpose.

Yet most brain-related activity has thus far focussed on specialized interests within individual disciplines and the continued reductionism of these myriad elements. As a result, brain science is still essentially a loose collection of diverse disciplines. In addition to neuroscience, psychology, neurology and psychiatry we are increasingly seeing contributions from the physical sciences, mathematics and information sciences. Numerous books attest to this diversity by presenting a wide, and at times bewildering range of perspectives on brain function, each with its own language and goals.

However, recent multidisciplinary efforts have provided the impetus to break down the boundaries and encourage a freer exchange of information across disciplines. This book is a consequence of these efforts: it is concerned with the manner in which all the brain's processes are interrelated, resulting in a system that is diverse yet unified, dynamical yet coherent; it is also concerned with the reasons for and consequences of dysfunctional processes.

> It predicts an emerging multidisciplinary field that might be called 'Integrative Neuroscience'

Another conviction motivating this book is the possible existence of a small number of key mechanisms and organizing principles underlying all of the brain's functions. It is true that for as long as some of these principles remain unknown, the ideal embodied by Integrative Neuroscience will be unfulfilled. However, the history of science is full of examples of fields that were transformed by ideas that were few in number and (in retrospect) simple. This book anticipates that sufficient facts about the brain already exist, that even now would significantly benefit from integration, and through which the organizational principles of the brain can be discerned. It may not be long before Integrative Neuroscience emerges as a distinct discipline that can lift us clear of the jungle of detail, and reveal new and broader perspectives of brain function.

The book chapters span the landscape of brain science to present core information from a range of disciplines. A casual glance through them may reinforce the impression of bewildering diversity. However in presenting core information across so many disciplines in neuroscience side-by-side, the intention is quite the opposite: it is hoped that the juxtaposition will help to promote in the reader an integration of ideas and the emergence of deeper and more general models.

We deliberately do not attempt in this book to present the latest exhaustive list of details about brain function. Nor do we provide idiosyncratic unified theories about the brain. Rather, we provide a perspective about the potential for integration, and offer a number of exemplars that may reflect future directions in this regard.

This big picture chapter serves to give a taste of what to expect from emerging integrations. It begins by outlining the nature of the problem. It then presents some seminal brain mechanisms across-scale. It then summarizes some of the more cited biological ideas concerning the organization of the whole brain as a system, and shows how a complementarity of these models should lead naturally to their integration. It is also suggested that simplification may be possible through merging of existing models as a first step. This overview encourages a model-based integration of information as the most expeditious way to capture the essence of the details in brain science. Finally, it examines possible disturbances in brain function.

## I. CONCEPTUAL INTEGRATION

It is a curious aspect of human brain science, which distinguishes it from possibly all other fields of science, that such an abundance of models coexist — despite the fact that they all seek to describe the one organ.

In most models of the brain, particular aspects of brain function are considered in isolation for example, anatomical structures, types of memory, particular time scales etc. The vast majority of the world's brain scientists focus on the microscopic scale and on specialized brain networks. This has led to an extraordinary information explosion that constitutes the essential building blocks underpinning all of brain science. However there are real limits on the extent to which mechanisms operating in a single neuron or specialized neural network can be scaled up into useful models of the whole brain, where phenomena result from the collective behavior of many interacting networks operating in parallel. The mechanism for harmonizing the contributing networks is simply lacking (or still controversial as in the case of 'Gamma' activity discussed in point 5 below).

There is also a sense that people building models at one particular scale often see that scale as the epicenter of brain function. We might all benefit from a larger context of the overall brain landscape. No model or scale of function yet holds a monopoly of wisdom. It is with this intention that this Big Picture section and the chapters in this book have been put together. This book seeks to illustrate the value of moving beyond any narrow scale or dichotomous (for example specialized versus holistic) approach, towards the foundation of a more integrated understanding of brain function.

This is all very well in theory. But in practice, there remains the very real problem of scale: which ranges from nanometres for genetics and the cell membranes of a single neuron, to millimeters for neural networks, and to centimeters at the scale of the whole brain. As well there are different degrees of abstraction: ranging from explicit biochemical and biophysical processes to concepts such as information processing, symbol manipulation and self-organization. As a result, we might appear to have in brain science a modern-day 'Tower of Babel' (see figure 1.1), built with a single purpose, but nevertheless doomed by jargon and misunderstandings between the various builders. That would indeed be a pity. Not only would the common goal of neuroscience not be reached, but researchers would be discouraged from venturing beyond their own specialty. This is precisely what seems to have occurred for some neuroscientists.

There is however an alternative possibility. Realistic integrations may now be a distinct possibility. As has been demonstrated in other complex dynamical systems: **simple rules may underlie complex behavior; different rules of function may occur at different scales and they may readily coexist.** For example, in the case of the weather, the daily and seasonal patterns are ultimately described by the gas laws. We clearly cannot expect there to be anything as simply expressed as the gas laws to explain all aspects of brain function. However if the history of science is any guide, we might expect sudden insights that shed new light widely over what is presently a fragmented field.

There are already plausible coexisting rules operating at each scale: genetic thresholds at the molecular scale, rate of firing at the single neuron scale, pattern of firing in neuronal networks, and feedforward–feedback processes at the whole brain level. There is an increasing number of insights into these interactions across-scale in the brain as a whole system. Helping to shape such integrative concepts of whole brain function, has also been a rising tide of interest in 'Complexity Theory' and the generation of new tools to explore the way complex systems like the brain organize themselves (for example Wiener, 1948; von Bertalanffy, 1968; Prigogine, 1989; Kauffman, 1993).

In summary, while we should be humble about our possible level of ignorance about how the brain has evolved and functions — a discernable outline is emerging of its possible integrated activities. In subsection II, a selection of 10 brain mechanisms and organizing principles, demonstrate different rules coexisting at different scales of function. The importance of models as the best way to distil the essence of an inordinate volume of details in neuroscience (and shed light on the mechanisms) is briefly outlined in subsection III. Some of the more cited biological models at the whole brain scale (across numerous academic disciplines) are then sketched out and integrated, to provide a preliminary outline of a dynamical working brain. The final subsection shows the potential to extend integrations about normal brain function, to functional instabilities in neurological and psychological disorders.

## II. TEN ACROSS-SCALE BRAIN MECHANISMS

If the overall flow of brain dynamics is to be brought together into more integrated models, it will require not just identification of the networks involved in cognition and behavior, but also elucidation of their mechanisms of action and organizing principles for unification.

Lets explore 10 selected mechanisms, that illustrate a possible flow of information processing in the brain (also see figure 1.18) from a bottom-up perspective (beginning from a single neuron).

1) The genetic material (constituting the DNA) in each of the brain's 10 billion neurons shapes the cascade of processes in each neuron. Inherited genetic thresholds modulate many neural functions, as evinced by The Genome Project and twin studies (Seligman, 1994).

2) Each neuron has multiple branch-like fibers (dendrites) which carry incoming signals to a neuron body, where action potentials are triggered that travel at constant speed down its axon to end branches, where the signal precipitates the release of stored brain chemicals (neurotransmitters) that traverse a tiny gap (synapse) to form the communication link that either excites or inhibits other neurons to which it is connected. All information transfer in the brain consists of these electrical and chemical processes. All of the brain's consequent functions are tied up with the shifting balance between this excitatory and inhibitory activity.

   Since the speed and size of the action potentials are constant, a critical communication code between neurons is the rate of firing, which averages 5 action potentials per second (Hz) in the cortex and typically ranges from 1–40 Hz, but increasingly faster in subcortical structures, spinal chord and sensory receptors (Kandel, Schwartz and Hassell, 1991; see models in Churchland and Segnowski, 1994 and chapter 5).

3) Multiple interconnected neurons constitute overlapping neural networks. Which subset of networks is active at any moment, is possibly selected according to competitive principles of natural selection that best achieve our adaptive needs at that moment (Edelman,1987). During critical early periods in brain development 'pruning' occurs of those networks that were not selected. There are a number of mechanisms that operate at this neural network scale. A key mechanism thought to constitute learning in the brain is reinforcement of connections between neurons that fire simultaneously ('neurons that fire together wire together'), and is known as Hebb's rule (Hebb, 1949). The corresponding biological process was discovered by Bliss and Lomo in 1973 and is called 'long term potentiation' or LTP.

4) A mechanism for feedforward synchronization at this network level is 'premodulatory coincidence', which is involved in anticipating the processing of environmental sensory events (Kandel and Hawkins, 1992). There is also a network mechanism of 'corollary discharges' in sensory-motor (and possibly other information processing), to allow for rapid correction of errors in the ongoing execution of motor activities (Sperry, 1950).

5) One mechanism by which the brain is thought to synchronize or orchestrate its network activities (in other words solve the 'binding problem') is known as 'Gamma phase synchrony'. It is proposed that networks activated at forty cycles per

second (Gamma) and in phase may underlie the coherence with which the brain integrates information (Singer, 1995). This mechanism is a testable exemplar of how the brain might achieve the trick of integrating its activities. 'The brain encodes information not just in the firing rates of individual neurons, but also in patterns of which groups work together' (Barinaga,1998). Gamma rhythms may be 'universal functional building blocks in the brain' (Basar-Eroglu *et al.*, 1996).

6) The whole brain is a highly interconnected system, functionally as well as anatomically. The coherence of its functions are dependant upon the interaction between specialized networks on the one hand (for example reflex, sensory, motor, language and face processing) and non-specialized more integrative networks on the other. Specialized reflex survival networks have evolved to detect potentially threatening situations and get us out of trouble as rapidly as possible. Survival networks therefore help us to avoid danger and minimize pain. The best known of these networks is the 'fight and flight' (hypothalamic-pituitary-adrenal) axis. Specialized sensory input networks (for elementary features of sound, touch, sight) are initially processed in specialized cortical areas in the back half of the brain. Specialized motor output (including speech) is executed towards the front of the brain (see figures 1.2–1.4). Details of these specialized networks can be found in the anatomy, sensory-motor and neurology chapters.

Related specialized networks in the body's autonomic nervous system may also be integral to higher brain functions and behavior. For example, Porges' (1998) 'Polyvagal Theory' proposes that the earliest autonomic nervous system was controlled by an unmyelinated visceral vagus (which is the tenth cranial nerve) that fostered immobilization behaviors in response to threat. A second stage evolved with the sympathetic nervous system allowing 'fight or flight'. A myelinated vagus, which is found only in mammals, regulates heart output and allows engagement and disengagement with the environment in a more flexible manner. This mammalian vagus is linked functionally to other cranial nerves that regulate social engagement via facial expression and speech. 'Parallelling this change in the neural control of the heart is an enhanced neural control of the face, larynx, and pharynx that enables complex facial gestures and vocalizations ... and greater regulation of behavior, especially social interactions' (Porges, 1998). Hence multiple stages in the evolution of a single cranial nerve are predicted to underlie diverse aspects of adaptive and social behavior.

7) The majority of the human brain is, however, non-specialized *association cortex*, involved in bringing together or integrating information and processing options. The trebling in human brain size over the past five million years has primarily been in this non-specialized association cortex (see chapter 3 for details). These association networks are integrally involved with the limbic system — which determines our goals, evaluates potential rewards and is seminal to processing memory and emotions. The limbic system is thought to generate our goals, and the cortex is involved in their detailed evaluation and implementation (Freeman, 1995).

The overall whole brain flow of specialized and association network processing is presented in subsection IV.

8) A conceptualization of the overall flow of the brain's chemistry remains contentious. The large number of receptor sub-types for many neurotransmitters

and the burgeoning number of discoveries of slower acting neuropeptides makes for a dazzling interplay of possible interactions at every level.

Nevertheless, there is some consensus of the importance of some of the major pathways. A good starting point in any brain model is sensory-motor processes. The basal ganglia (with dopamine having an inhibitory effect on these neurons) is thought to exert a predominately inhibitory function on the thalamus, leading to a reduced transmission of sensory information to the cerebral cortex as well as a concomitant reduction of arousal (which is controlled by the ascending reticular formation and operates in close linkage to the sensory input). 'The purpose of the thalamic filter mechanism should be to reduce the transfer of sensory input to the cortex in a selective manner, such as to protect the cortex from an overload of irrelevant stimuli and to allow the cortex to focus on novel and relevant information' (Carlsson *et al.,* 1997).

More broadly, the primary excitatory neurotransmitter in the cortex is glutamate and its major inhibitory neurotransmitter is GABA. Four neurochemicals manufactured in subcortical regions of the brain, distributed widely and considered seminal to a number of brain dysfunctions are: dopamine (from the substantia nigra and ventral tegmental area) which is essential to the functions of the basal ganglia; acetylcholine (from the nucleus basalis of Meynert) and involved in memory; norepinephrine (from the locus ceruleus) and serotonin (from the raphe nucleus) which modulate more global aspects of information processing including aspects of arousal and emotion respectively (for further details see chapter 6, The Brain's Chemistry, including the potential role of longer acting neuropeptides). The overall balance of these excitatory and inhibitory neurochemical activities underlie all brain processes and the extent of its overall stability.

9) At a more conceptual level, different networks are engaged according to situation and task demands. Brainstem and limbic system survival networks are involved in 'Automatic' crude and rapid processing (LeDoux, 1994). Incoming information is routinely compared with stored information ('comparator' networks) and a mismatch results in 'controlled' more detailed and slower processing preferentially involving non-specialized networks (Schneider and Shiffrin, 1997; Ohman, 1979; Gray, 1995).

10) What about integrations of all mechanisms of brain function? A unifying theory of the brain still seems a long way off, so such organizing principles will have to start by bringing together existing models to achieve unifying perspectives. The next subsection demonstrates an integration of whole brain models. But new approaches to conceptualizing parallel interacting network functions across scale are also required.

A speculative approach by Gordon and Rennie (1999) serves to demonstrate that such common perspectives are possible, at least in a rudimentary form. Our model has an evolving network focus ('purposeful dynamical networks' or PDNs) as the brain's flexible functional units. PDNs transform inputs, evolve their synaptic strengths, link up with other networks to facilitate the parallel feedforward-feedback processes that underlie the brain's ongoing decision making activity, whilst retaining an overall adaptive functional stability. The model builds on the integration of overall information flow in the brain (reflected in figure 1.15), is also consistent with the overall pattern of excitation–inhibitory neurochemical flow (outlined in point 8 above), and melds with a number of other theories: Edelman's notion (1987) of network adaptation through natural selection; Wright's picture of spreading activation with linear subcortical modulation (chapter 10a); Freeman's ideas about the role of the goal setting limbic system in modulating the strength and focus of non-linear cortical activity (chapter 10b); Sokolov and Gray's notions of 'comparator' processes (of incoming versus stored information) with mismatches or matches of such comparisons — resulting in new explorative, or ongoing continuity of functional trajectories respectively.

The key point of the PDN model is to demonstrate one means to break away from any specific spatial, temporal or scale-centric view of the brain: be that specialized, modular, reflex, or global. A PDN type model also brings together as a rather natural extension, four overall organizing principles that achieve this perspective:

a) Adaptation. Brain function is self-organized to adapt continuously to its current environment.

b) Dynamics. A limitation of many current models is that the temporal dynamic is omitted. *Timing* is the key to adapting to a continually changing environment.

c) Anticipation. Feedforward processing is integral to harm avoidance and there are a number of simple insights, such as the continuity of our visual worlds, that suggest the importance of ongoing feedforward processing. Most information processing also leads to expectations. Yet few models address the problem of choosing between competing goals. Figure 1.16 illustrates the flavor of such a dynamical feedforward adaptive model.

d) Dimensional. A myriad models either focus on 'specialized' networks or on distributed 'holistic' models. In reality they readily coexist along a continuum, which is differentially activated according to specific situations and task demands. For example, sensory-motor reflex processes would engage multiple specialized networks. Seeking reward on the other hand, might involve evaluative networks and a number of less differentiated 'association' networks for processing multiple options, possibly more autonomously from the sensorium.

Following a *purpose* of the brain's function, might provide a natural window through which to explore integrative mechanisms across scale.

## III. THE IMPORTANCE OF MODELS

Models are the only way to capture the essence of the mass of details, determine their possible interactions in an explicit manner, elucidate possible mechanisms and compare this theory with real data. New findings are of little value if they lead recursively back to validating our preconceptions. In a brain with 10 billion neurons highly interconnected, only models can realistically be expected to winnow out the significant network interactions in the brain.

The goodness of fit between model-based theoretical predictions about brain function and real brain data, is the final arbiter of the strength or weakness of a model. *Essentially, the goodness of fit between theoretical models and real data is the basis of all science.* The more quantitative (numerical) this comparative process is (between theory and data), the better. Box and arrow models and statistical correlations play an important role in the modeling process, but are not as powerful as numerical simulations (mathematical simulations of explicit interactions fitted to quantified data).

There is now the motive, the means and the opportunity to begin to integrate some of the myriad outcomes in neuroscience:

i)   Mechanisms across scale (from single neurons to behavior).
ii)  Specialized networks including sensory input, motor output (including language), survival reflex, comparator decision making networks.
iii) Whole brain feedforward–feedback processes.
iv)  Adaptive, dynamical, anticipatory, dimensional perspectives and brain–body interactions.

Mathematics and computers offer a means of describing concisely many of these processes. The consequences of all these processes operating in concert, which can otherwise only be guessed, are made explicit by mathematical analysis or numerical simulations.

Modeling brain function has also long been the inspiration for the fields of Artificial Neural Networks (ANNs) and Artificial Intelligence (AI), in which the emphasis is on mimicking intelligent behavior with varying degrees of biological faithfulness (see chapter on computer models). Many extraordinary ANN models emulating specialized sensory, motor, language and face recognition processes have been achieved. ANN models are increasingly incorporating biological parameters (Amit, 1989) and exploring realistic structural and functional connectivities of brain function (for example Tononi *et al.,* 1994). Recent developments in AI also include a system level interest in both symbolic and connectionist solutions to the problems faced by autonomous agents: which include acquisition of skills, choosing between competing routes to a goal, and prioritizing several simultaneous goals. 'Competence modules' interactively are programmed to be 'goal oriented, opportunistic, persistent, able to plan ahead, robust, reactive and fast' (Franklin, 1998). Such tasks are similar to those discussed in the context of adapting human brains.

Not so well known but of seminal relevance to brain models, is the area becoming known as Brain Dynamics. Here mathematics and computers are used too, but in a way that is a compromise between the small scales implied by ANNs and the abstractions dealt with in AI, and fits naturally with the types of measurements that can be made and the global circuitry that can be discerned in the brain.

Walter Freeman's model (chapter 10b on brain chaos and intentionality) is the result of meticulous measurements, particularly of the olfactory tract. He has managed to capture much of the observed dynamics of electrical activity in a set of equations, and to perform numerical simulations using these equations, that match observations. It emerges that the stability of the brain seems to depend on subtle balances, too much or too little order may prove disastrous. Some of the brain's interactions may be linear, others highly non-linear and discontinuous. 'The brain has been selected to become so unstable that the smallest effect can lead to the formation of order' (Prigogine in Briggs and Peat, 1989). Langton (1990) suggests that the enhanced flexibility of function found at the 'edge of chaos' has been a naturally selected dimension of the brain. Linear and non-linear brain processes are therefore partial descriptions of an underlying unified brain process, and their relative predominance would seem to be situation and task dependent.

In chapter 10a, Jim Wright describes another numerically explicit model exemplifying brain dynamics. As in Freeman's model it makes no attempt to reproduce the firing patterns of individual neurons. Instead it aims to model the collective behavior of large ensembles of neurons, matching the scale with which we can record activity easily in humans (using computerized brain imaging technologies described in chapter 15). The parameters in the model are biologically realistic, including details about how neurons are interconnected over short and longer range, the rate of firing of network activity, the speed of conduction of electrical activity in the brain (6–9 metres per second), the effect of various excitatory or inhibitory neurotransmitters, modulation of brainstem arousal on overall brain stability. Although it models overall electrical brain function, there are strong reasons to believe that there is a direct connection with cognition (see chapter 10a). There is a considerable benefit that comes from modeling on this larger scale, since non-linearities and chaos seem to become less apparent in some states of brain activity. Consequently this model can be largely linear in the first instance and numerical simulations can be carried out relatively easily. Non-linear processes can then be explored from this frame of reference.

Recent contributions from Peter Robinson and Chris Rennie to the Jim Wright model (see references in chapter 10a), have explored it's characteristics analytically and achieved unprecedented solutions. A very rudimentary whole brain model has been successfully achieved. In other words, the simulations (based on anatomical and physiological parameters) have been found to fit remarkably well with real human data, including brain function at rest, sleep, various aspects of decision making and a raft of clinical data. Individual subject's data can now be fitted to the simulation and interpreted with respect to the mechanisms and parameters in the model. The numerical and mathematical tractability of this model has the attraction that it is open to the addition of anatomical and functional details. Researchers from other disciplines involved in brain science will be able to add new mechanisms and parameters: for example more anatomical details, other neurochemical or metabolic interactions, the way in which the speed of decision making might be modulated etc. These contributors will then readily see how well the numerical simulation fits with their real data.

This particular model may turn out to be wrong or insufficient. But it has demonstrated that integrating key aspects of brain anatomy, physiology and chemistry into a realistic whole brain model, is achievable. This sort of numerical simulation demonstrates that it is now possible to model integrative aspects of overall brain dynamics.

## IV. INTEGRATION OF MODELS OF THE WHOLE BRAIN

To pursue the overall problem of the brain trying to understand itself is to confront the ultimate complexity. Every movement we make, fact we remember, or subjective state that we feel, appears so effortless and instantaneous. Yet this is achieved by 10 billion neurons, most of which appear to be firing at random, and all of which are intricate in structure and unique in their interconnections. Clearly there are powerful but poorly understood forces of self-organization that integrate this neuronal activity, giving it meaning and purpose, across microscopic, millimetric and whole brain scales of function. Current models at the whole brain scale of function only hint at these critical rules of self-organization, yet are still the best starting point in the quest. *They provide surprisingly complementary insights.*

Let us explore this complementarity, by considering some of the most cited models that have a biological focus pertaining to whole brain function. Each selected model from each discipline is briefly described in the text below (the reference list provides a source for a more detailed explanation of this information). Visual distillations of these models are presented in figures 1.5–1.13. An exemplar integration of these models is then provided and distilled in figures 1.14–1.15. *Figures 1.1–1.17 courtesy of Evian Gordon www.brainart.com.au (1.18 courtesy of Evian Gordon and Edyta Delfel).*

The summary of selected whole brain models begins with the best known neuropsychological model, that of Luria (1973), in which interaction among three

## "Tower of Babel" across scales and disciplines

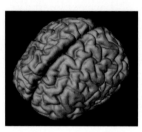

WHOLE BRAIN:

Adaptation
Self organisation
Feedforward
Feedback
Temporospatial
Reafference copies
Attention
Info processing
Defences

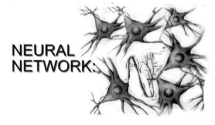

NEURAL NETWORK:

Hebbian simultaneity
Synaptic mod
Receptor subtypes
LTP, LTD, Gamma
Habituation
Sensitization
ANNs

SINGLE NEURON:

Action potentials
Rate of firing
Second messengers
cfos
genes

Figure 1.1

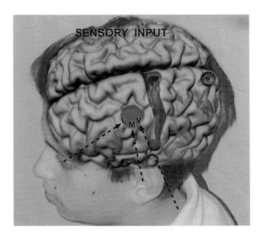

Figure 1.2

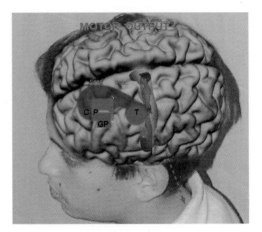

Figure 1.3

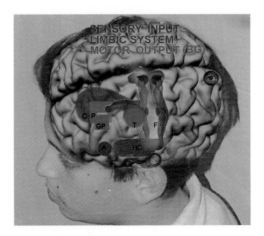

Figure 1.4

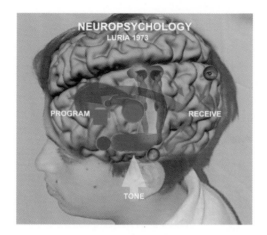

Figure 1.5

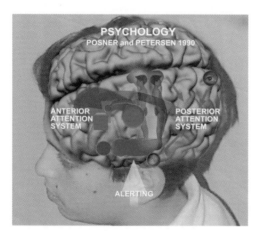

Figure 1.6

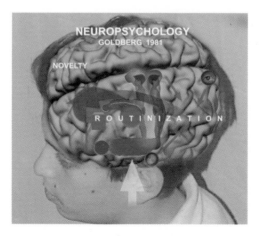

Figure 1.7

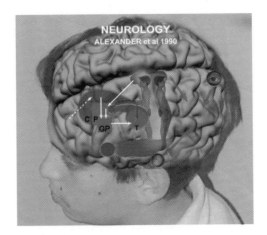

Figure 1.8

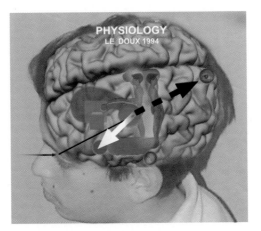

Figure 1.9

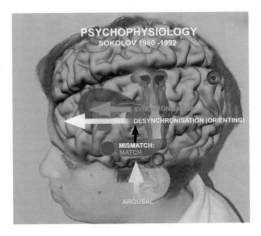

Figure 1.10

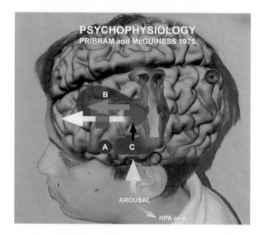

Figure 1.11

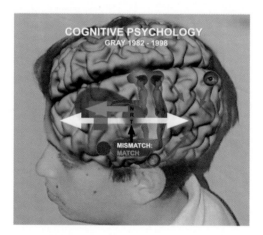

Figure 1.12

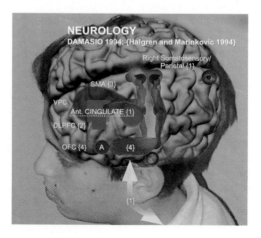

Figure 1.13

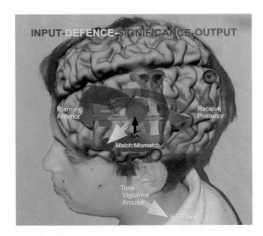

Figure 1.14

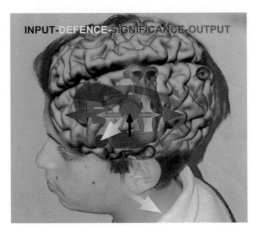

Figure 1.15

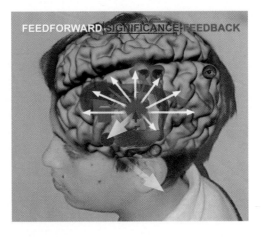

Figure 1.16

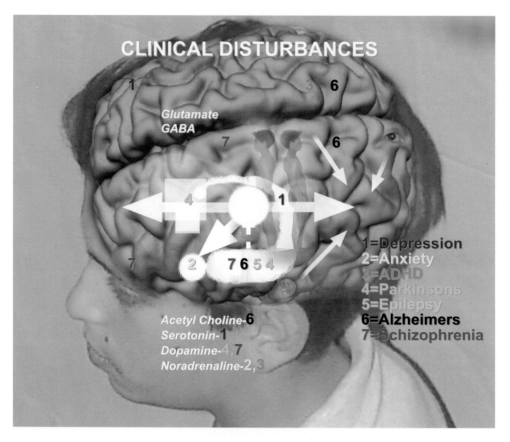

**Figure 1.17**

principal functional units ('Receive; Verify; Tone') was considered for any type of mental activity (figure 1.5). The specialized 'Receive' input unit and the 'Verify' output unit, are modulated by a more distributed 'Tone' unit that adjusts the brain's overall functional level of arousal. Environmental stimuli are detected by receptors and transmitted via specific nuclei in the thalamus (see anatomy and sensory-motor chapters for details), to specialized primary cortical areas (mauve areas in the figure) situated towards the back of the brain. Luria proposed hierarchical processing (from specialized primary cortex to secondary to tertiary/association cortex), reflecting an increased integration of the elements of an environmental stimulus (mauve arrows in the figure) and a progressive lateralization of preferential hemisphere activity in processing this information. One of the brain's superhighways (the superior longitudinal fasciculis shown as the large mauve arrow in the figure) then transfers integrated elements of the information from the association cortex at the back of the brain, to the association cortex in the frontal networks, where the processing is 'Verified' and 'Planning' is made to execute appropriate motor responses via the primary motor cortex (orange strip in the figure). The 'Tone' networks from the brainstem act as a global 'nerve net' to ensure an increased level of function in processing novel information and reduced activity for more familiar inputs. The frontal lobe is considered the planning/verification/abstraction

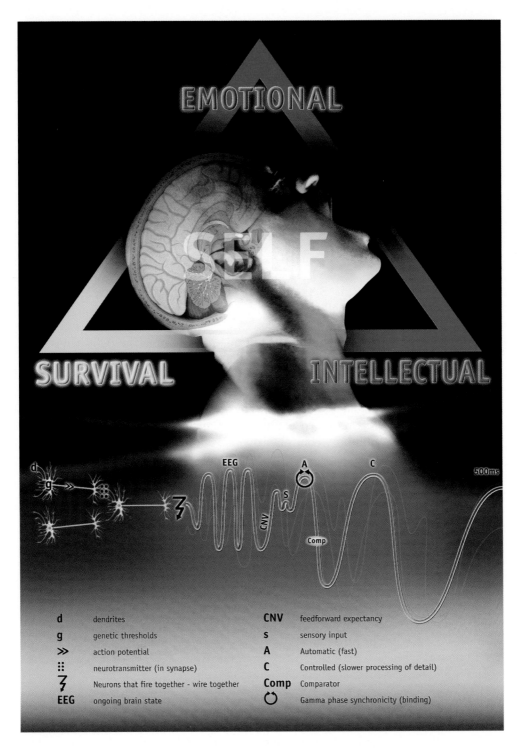

Figure 1.18

'superstructure above all parts of the cortex'. Whilst Luria proposed this rather unambiguous organization of information flow, this was by no means a simplistic localizationist model, but a dynamical one. He emphasized that higher mental functions are not static in the brain and 'our fundamental task is not to localize higher human psychological processes in limited areas of cortex, but to ascertain by careful analysis which groups of concertedly working zones of the brain are responsible for performing complementary mental activity'.

Luria's model provides a useful starting point for integrating overall brain function. It has been used extensively in the field of neuropsychology to explore lesion-behavior relationships, and is also a useful frame of reference with which to interpret findings from brain imaging technologies (chapter 15). A number of models have extended this basic outline, to explore fundamental aspects of brain function including attention, specific aspects of information processing and arousal. Some of the more cited of these models are briefly described below.

Posner and Petersen's (1990) model of attention (figure 1.6) is extensively cited in the psychology literature. It follows a similar pattern to Luria's model with a 'Posterior Attentional System' (consisting of the posterior parietal region, the pulvinar nucleus of the thalamus and superior colliculus) associated with orienting to new information (within 150 milliseconds), and associated with lowering thresholds for a motor response (modulated by the neurotransmitter norepinephrine). The right hemisphere is biased toward global processing (low spatial frequencies) and the left for local processing (of high spatial frequencies). Further processing of the detected target stimulus is then 'handed off' (large mauve arrow in the figure) to the executive 'Anterior Attentional System' (consisting of the dorsolateral prefrontal cortex and the anterior cingulate) for coordination of analyzing functions and becomes particularly active in tasks where inhibitory control or divided attention are critical (this network being modulated by both dopamine and norepinephrine). Their 'Vigilance' network (yellow arrow in the figure) is similar to Luria's 'Tone', namely an arousal system that maintains an adaptively appropriate alerting state, and according to this model preferentially involves the right hemisphere and modulation by norepinephrine.

Goldberg's idea of a dynamic relationship between cortical regions in the process of learning, represents a paradigmatic shift in the way neuropsychologists think about the brain (figure 1.7). It lays the groundwork to shed light on the explicit functional neuroanatomy of information processing and learning. He suggests that the dynamic aspects of brain organization at the level of cognition are best explored with the current methods of functional neuroimaging. The distinction between cognitive 'novelty' (processed in the right prefrontal cortex) and 'routinization' with more familiar information (processed preferentially in the left hemisphere) is central to the description of cortical functional spatio-temporal dynamics. The expression of this general principle of hemispheric specialization in the posterior association cortex and in the frontal lobes has been extensively explored (Goldberg, 1990, Goldberg *et al.*, 1994). The 'dynamic' neuropsychology developed by Goldberg represents a direct extension of, and elaboration upon, the classic notions developed by Alexander Luria.

There are also in the literature a number of commonly cited models that explore more specific brain activities that are seminal to whole brain cognition, including motor output, fear orienting, arousal, novel/familiar information processing and emotions.

The brain's sensory-motor networks are focussed upon in the discipline of neurology. Alexander, Crutcher and De Long (1990) have used anterograde and retrograde stains to delineate networks in a range of motor-related activities. They show a parallel funneling of multiple feedforward inputs from cortical areas to sub-cortical basal ganglia structures (green structures in figure 1.8) which converge (white and pink arrows in the figure) on the thalamus — whose feedback output is centered on one part of the frontal lobe (and on the anterior cingulate in the case of the 'limbic' pink network). As shown in the figure the 'Motor' (dorsal) networks consist of pathways that flow through the putamen (P), whereas the 'Oculomotor' and 'Prefrontal' (dorsolateral and orbitofrontal cortex) networks flow to the caudate (C) nucleus. A 'limbic' (Ventral in pink) pathway includes the nucleus accumbens which connects to the anterior cingulate before communicating with the globus pallidus (GP) and thalamus (T). The model also highlights the interactions among the cortex's primary excitatory neurotransmitter (glutamate) and inhibitory dopamine in the basal ganglia (with dopamine generated from the substantia nigra in the dorsal motor and ventral tegmental area in the limbic pathway). There are also 'direct' and 'indirect' pathways (engaging the globus pallidus and the sub-thalamic nucleus respectively) which 'open and close' the brain's core sensory-motor flow of information via these central thalamus-cortical networks.

The LeDoux (1997) model emerged from physiological research into fear conditioning. He demonstrated that fearful situations are initially processed rapidly 'Automatically' and subcortically in the amygdala (white arrow in figure 1.9) via the midbrain and thalamus (reflected in the thin black arrow in the figure) — to keep the animal out of trouble. It is the lateral nucleus of the amygdala that receives the inputs from the thalamus (with excitatory glutamate activation) and connects to the central nucleus of the amygdala, which initiates expression of a spectrum of emotional responses including heart rate, respiratory rate and activation of the hypothalamic-pituitary-adrenal 'fight or flight' axis (small yellow arrow in figure 1.11). In parallel, a slower 'Verification' and more detailed processing (black broken arrow in figure 1.9) is undertaken by the cortex. Spatial information processing and the 'context' of the emotion involves the subiculum of the hippocampus. This model exemplifies the fact that if the situation demands survival based processing, the earliest network activations may not even involve the cortex, which is subsequently activated for more detailed processing. This model highlights how brain functions may be situation and task specific, without resorting to the need for 'superstructures' in the brain that necessarily oversee all of its processes.

Sokolov (1960–1990) also proposed a model of brain function that was situation and task specific. Interacting reflexes of 'Defence', 'Orienting' (to novel stimuli) and 'Adaptation' (conditioning) were considered to underlie all aspects of brain function, to form 'Neuro-Humeral Control Systems'. Habituation (response decrement of the 'orienting' reflex with repetition of novel stimuli) was the most ubiquitous form of learning in the brain and is currently incorporated into many models of learning. Sokolov presumed a 'comparator' process (neuronal model) that compared incoming environmental information with stored experiences of information. 'Mismatches' resulted in orienting reflexes (automatic head and eye movement, increases in a range of body functions) in response to novel or significant information. Sokolov predicted that comparator mismatches in the hippocampus (due to novel information), would result in thalamic signals to desynchronize cortical activity (white arrow in figure 1.10) and increase brainstem arousal (yellow arrow in the figure). More synchronized processing (green arrow in the figure) and decreased brainstem arousal was associated with more familiar information. The location of the comparator match or mismatch processing has changed from the hippocampus to the cortex, to multiple sites in novelty detecting cells with the development of his model, but the principles have remained the same.

The model of Pribram and McGuiness (1975) focuses on arousal networks linked to brainstem (reticular) activity and the 'fight and flight' network (hypothalamus-pituitary-adrenal axis represented by the yellow arrows in figure 1.11). These arousal (C) networks are particularly associated with the amygdala (A in the figure). The hippocampal 'Effort' Networks (C) are proposed to coordinate interactions between 'Arousal' (A) and basal ganglia 'Activation' (B in the figure) networks (further reviewed by Boucsein, 1992).

Gray (1982–1998) explored from the perspective of cognitive psychology, the consequences of a core matching or mismatching process between 'the current state of the organism's perceptual world and a predicted state'. He proposed that this 'comparator' process occurred in the subiculum of the hippocampus every tenth of a second, and the output formed the core contents of consciousness. He focussed upon the networks engaged as a consequence of 'novel' mismatch (white arrow in figure 1.12) or 'familiar' matched information (green arrow in the figure). Novel stimuli also engage a 'Behavioral Inhibitory System' (Gray 1982). From the subiculum (and entorhinal cortex) the meaning of the stimulus (its extent of novelty or familiarity) activates the nucleus accumbens (a key structure of the limbic network in the Alexander model and part of the pink network in the figure), which is proposed by Gray as the gateway to the basal ganglia and motor output (the 'Behavioral Activation System' and the green structures in the figure). *Inputs to the comparator can therefore evoke entirely different network trajectories.* A *match* signal to the nucleus accumbens permits the input to the accumbens from the amygdala to trigger the initiation of the next step (green arrow) in the motor program (green structures). Whereas a *mismatch* decision is passed on in two ways: as a message to the cingulate cortex where it brings the current motor program to a halt; and as a generalized message to the nucleus accumbens, which triggers exploratory behavior (white arrows). In either case (match or mismatch) the nucleus accumbens and reticular nucleus of the thalamus (NRT in the figure), are involved and 'a message is transmitted back to memory stores in the temporal lobe, indicating either that a stored associative regularity has been confirmed or that memory stores need to be updated' (1995).

All of these models reflect to a different extent the interplay between brain and body interactions. Damasio's (1994) model deliberately highlights the possibility that our lowest and highest levels of cognition and decision making include our emotions, which are integrally linked to body functions. 'The mind is embodied, not just embrained.' LeDoux's model focussed on the role of fear in our survival and he proposed that 'emotions or feelings are conscious products of unconscious processes ... triggering systems of behavioral adaptation that have been preserved by evolution'. Damasio's more general 'Somatic Marker Hypothesis' sees emotion as the bridge across levels of function and proposes body networks that bias our central decision making to avoid pain. Innate body networks, shaped by previous experience, act as an automatic 'alarm' to modulate the brain's decision making networks against selecting the option leading to the worst adaptive outcome. Whilst these processes are synchronized by timing (with higher frequencies reflecting pleasant sensations and lower frequencies pain) his lesion studies in neurology have focussed on the spatial networks (summarized in figure 1.13) thought to underlie emotions (and their interrelationship with 'reasoning'). The networks include the dorsolateral prefrontal cortex (DLPFC in the figure) associated with working memory and reasoning; the networks preferentially involved in emotions are the ventromedial prefrontal cortex (VPC), the right somatosensory parietal region (including the insula which is not shown in the figure), motor networks (orange primary motor strip and supplementary motor area or SMA), the amygdala (A) and the anterior cingulate cortex (shown on the cortex in the figure, but the broken line indicates it is a subcortical structure). He suggested that the anterior cingulate was the 'fountainhead' for interactions among emotional, attentional and working memory processes.

Models of emotion are deliberately included in this Big Picture section to illustrate that plausible network activity (based on reasonable empirical evidence) is already being postulated to underlie even some of our most complex levels of cognition. Speculative models drawing on such neuroscience, have also begun to postulate preferential networks that regulate our emotions and behavior, particularly networks in the right orbitofrontal cortex, which are shaped by experience between mother and child during early critical development periods, and modulated by social forces (which shift our early adaptations 'from a pleasure to a reality principle'). Inhibitory orbitofrontal cortex activity onto deeper brain structures is thought to facilitate the temporal organization of our emotional behavior (Schore 1994).

The final mainstream model (also presented in figure 1.13) is that of Halgren and Marinkovic (1994), since they focus on the temporal dimension of processing emotion. They propose that emotion 'seems to function to provide a directed integration of the entire organism'. They dissect the neural basis of emotion according to different time scales. Four overlapping stages are distinguished from psychophysiological research on brain and body functions. Networks associated with each of the four stages are numbered: 1.Orienting (rapid automatic emotional appraisal associated with the cingulate, posterior parietal and hypothalamic networks); 2. Event integration (slower detailed 'emotional coloring' or meaning, in association cortex with specific emotions additionally processed in specific networks, for example, fear in the amygdala, pleasant emotions in the septum, disgust in the insula (Phillips et al., 1997); 3. Response Selection (motor networks); 4. Context (sustained 'mood'/temperament with a longer time scale shaping functions via orbitofrontal cortex [OFC], hippocampus and hormonal activity).

The Luria, Posner, Goldberg, Alexander, LeDoux, Sokolov, Pribram, Gray, Damasio and Halgren models each demonstrate different but fundamental aspects of adaptive brain function. Yet they retain some familiarity and natural complementarity. Can these models be integrated? Figures 1.14 and 1.15 demonstrate that this is readily achievable. If we strip away the jargon of each discipline and bring the models together, we see the tentative outline of an adaptive whole brain model — that is, specialized yet integrated, dynamic yet potentially stable and self-organizing.

Figure 1.14 serves as an exemplar of Integrative Neuroscience in bringing together the above complementary mainstream models. It shows the complementarity of adaptive Input-Evaluative-Output functions and a 'nuts and bolts' overlap of specific networks among the models. The figure shows interacting sensory input (mauve), survival (yellow), limbic decision between novel and familiar (pink), and output motor (green) networks. The figure incorporates the basic flow of information outlined by Luria: the mauve arrows show the Receive networks towards the back of the brain integrate elements of information, and the result is transferred to the front of the brain for Planning and execution of motor responses with modulation by the Tone networks. The similarity of this flow to Posner's Anterior, Posterior and Vigilance Attentional networks and Pribram's Arousal model has been outlined in the text. Alexander's model (green structures) highlights the motor output processes, and the figure reflects how the evaluative outcomes in the limbic system are transferred to motor networks, as well as the opening or closing of the thalamo-cortical flow of information. The amygdala (in the LeDoux and Damasio models) is integral to survival and emotions and is also associated with arousal in the Pribram and McGuinness model through activating the hypothalamo-pituitary-adrenal axis: the 'fight or flight' network that is essential to survival as well as being involved in stress and immunological functions. The pink 'evaluative' limbic networks are involved in the 'comparator' processing of the outcomes to predicted expectations in the Sokolov and Gray models. Both models (Sokolov and Gray) show a different trajectory for hippocampal 'mismatch' of novel information (pink arrow of novel information) versus 'matched' familiar (green arrow) processing, which would result in ongoing motor processes. Gray proposes that hippocampal evaluative outputs (black arrow) flow to the nucleus accumbens (NA in the figure which is also a key structure of the limbic network in Alexander's model). A match (green arrow) continues the motor program, whereas a mismatch triggers exploratory behavior (from the nucleus accumbens to the anterior cingulate).

Figure 1.15 emphasizes these processing flows. Ongoing sensory-motor information is processed posteriorly to anteriorly, with core decision making about stimulus significance occurring as matching or mismatching (of external and internal signals by the limbic system), with modulations by brainstem and cortical networks. Threatening information is processed crudely and rapidly, detailed processing is undertaken more slowly. *The discernable outline of a dynamical working brain emerges from the integration of these mainstream models across disciplines.*

The more integrated that such biological models become, the more likely they will be able to be extended further to models of behavior (including dimensions such as IQ/personality/creativity etc). Figure 1.18 provides a rough landscape view of The Big Picture interrelationships among aspects of *Behavior–Networks–Mechanisms* (the top black level in the figure shows behavior; the middle blue level reflects the neural networks that subserve these behaviors; the bottom grey level lists some of the mechanisms that operate these neural networks). It also highlights the ongoing bottom-up and top-down interactions among all the brain's processes. The processes and mechanisms in this figure are by no means meant to be exhaustive. Rather, they serve to outline the coexistence among these levels of function — that ultimately require detailed integration.

Aspects of behavioral models already embody elements that might link to the other levels. Three dimensions of behavior are outlined in this figure (Survival; Intellectual; Emotional) — each is subserved by diverse brain networks. The *Survival* 'fight or flight' networks have already been outlined. The *Intellectual* dimension is primarily reflected in practical resourcefulness, preferentially involving the association cortex. A somewhat controversial index of this capacity is IQ. These measures have been found to be predictive (to a varying degree in different circumstances) only for low stress situations and have little correlation with human creativity. Nevertheless, this concept still attracts considerable interest. The literature is split according to whether a general 'g' factor regulates a person's rate of learning and reflects their ability to deal with cognitive complexity (Gottfredson 1998), or whether 'multiple intelligences' exist (Gardner 1998) that more appropriately reflect the person's distinct strengths and weaknesses. Logical-mathematical, linguistic, musical, spatial, body-kinaesthetic, interpersonal and intrapersonal dimensions have been proposed. 'Emotional intelligence' (Goleman 1996) is a combination of the intrapersonal and interpersonal aspects of Howard Gardner's multiple intelligence model. There is the suggestion that multiple intelligence models inadvertently incorporate dimensions of personality.

Models of personality exist that attempt to link to biological findings. Cloninger's (1993; 1994) biosocial theory of personality for example, proposes four core dimensions of temperament: novelty seeking, harm avoidance, reward dependence, and persistence. They are 'independently heritable, manifest early in life and involve preconceptual biases in perceptual memory and habit formation'. These dimensions have been found to be related to different brain networks. For example, novelty seeking is related to behavioral activation and dopamine; harm avoidance is associated with behavioral inhibition and serotonin; and reward dependence is related to behavioral maintenance and norepinephrine. Networks such as these (novelty, avoidance and reward) have been explored to a considerable extent in a number of the models summarized in figures 1.5–1.13, highlighting the potential for further Integrative Neuroscience across diverse arenas.

Much less is understood about our *Emotional* soothing needs. The limbic system is thought to be central to our evaluation of the significance of information, emotional reward based behavior and self-soothing needs. Models of emotion and its regulation now abound in the literature. There are many other possible variables and interactions including developmental, cultural and experiential that require longitudinal assessment. An integrative multidisciplinary approach will, I hope, help to determine an appropriate critical analysis of the modulations of such processes on brain dynamics.

But there should also be caution about premature integrations, which are unlikely to be substantiated, without significant advances in Integrative Neuroscience in bringing together some core specialized and behavioral dimensions of brain function. Until basic conceptual frameworks are agreed upon, research efforts concerning more elusive concepts such as creativity, qualia (subjective feeling states), core soothing needs, consciousness and Self, might best be directed towards sorting the solid from the woolly pre-modeling bits of data. The more solid of these findings can then be explored as further building blocks, in emerging integrations of models of the human brain.

## V. WHAT MIGHT GO WRONG WITH THESE BRAIN FUNCTIONS

Integrations such as those in the previous sections, could provide a platform for a clinical perspective that highlights how, in a dynamical system, even slight imbalances in network function may result in severe disturbances of information flow and behavior. Each brain disorder is an experiment of nature, with unique insights into brain function that also highlight opportunities for intervention.

Disturbances in the brain can be structural (such as tumors or strokes that are readily detected with CT and MRI brain imaging technologies outlined in chapter 15). But the majority of disturbances in brain function are more subtle, resulting in imbalances of brain chemistry from a variety of sources. With 10 billion highly interconnected neurons and about 200 neurochemicals (each with multiple receptors which may have different functions) there is a huge range of possible imbalances, and corresponding symptoms. Many mainstream clinical models (psychological, neurological and psychiatric) focus on such neurochemical imbalances, because a number of clinical disorders can be at least partially treated with medication.

The overall flows of brain chemistry have been sketched: the basal ganglia (with brainstem dopamine) exert an inhibitory effect on the thalamus, leading to a reduced transmission of sensory information to the cerebral cortex — allowing it to focus on novel and inattend to irrelevant information. The primary excitatory neurotransmitter in the cortex is glutamate and its major inhibitory activity is GABA. Three other brainstem neurochemical modulators of cortical activity are acetylcholine, norepinephrine and serotonin. Longer acting neuropeptides also modulate many aspects of brain function and behavior.

Figure 1.17 shows a summary of some of the postulated neurochemical disturbances in common clinical disorders. The figure lists some of the chemical disturbances (on the left) found in these disorders (listed on the right). The figure also points to some of their dysfunctions (briefly summarized below with selected review references).

From this figure the reader will see the natural overlap and potential to link these clinical models to emerging mutidiciplinary integrations of the brain, outlined in the previous section.

In depression, increased limbic system activity, decreased right frontal lobe activity, dysregulation in the basal ganglia and anterior cingulate, are coupled with a shutdown of serotonin, norepinephrine and dopamine (George *et al.,* 1998; Mayberg, 1997). Anxiety Disorders have been attributed to disturbances in GABA, norepinephrine and amygdala networks, which also result in an inappropriate activation of the 'fight or flight' hypothalamic-pituitary-adrenal axis (Charney and Deutch, 1996). In Attention Deficit Hyperactivity Disorder, core disturbances in norepinephrine and dopamine involving frontal and possibly right hemisphere networks are thought to lead to disturbances in attention, arousal, hyperactivity and impulsivity (Pliszka *et al.,* 1996; Barkley, 1998). Parkinson's Disease has traditionally focussed on disorders of movement and well-established dopamine disturbances (in the substantia nigra), but recent findings of hippocampal atrophy have shifted attention to abnormalities in limbic–cortical cognitive processing networks. In epilepsy, hippocampal scarring has been associated with memory and other disturbances in information processing. Alzheimer's Dementia has been associated with acetyl choline and disturbances in the hippocampus (associated with memory), as well as decreased activity in the temporal and parietal regions early in the disorder and frontal networks in the later stages (Adams *et al.*, 1997). Postulated abnormalities in schizophrenia include increased dopamine, serotonin (and more recently decreased glutamate); disturbances in inhibition particularly in the early stages of information processing; mismatches (and misattributions of irrelevant information) in the hippocampus; decreased frontal network functions possibly more so in the left hemisphere, and thalamic and cerebellar abnormalities — all consistent with a distributed network failure of integrative information processing (Gray, 1998; Andreasen, 1997; Carlsson and Carlsson, 1990).

Multidisciplinary more integrative research efforts are gaining momentum in these disorders. Shared bigger picture views of the patterns of disturbance are also shaping the direction of research efforts, in a hands-on way. As one example, the potential impact of the notion that *simple rules of brain function may underlie complex patterns of behavior,* allows clear alternative possibilities to be explored:

1) In ADHD a core disturbance may be inadequate processing of novel stimuli, whereas patients with schizophrenia seem to have problems in selectively inhibiting irrelevant information. Behavioral or pharmacological treatments that magnify novel information or dampen processing of irrelevant information, will have very different results in such disorders, if this distinctive pattern of disturbances proves to be true.

2) Some patients with Parkinsons Disease cannot walk, but can dance. Such subtleties of motor programming demand a more holistic approach to the pharmacological and behavioral treatment of such disorders, than simple supplementation of dopamine.

3) 'Neurons that fire together — wire together, making it more likely that they will fire together in future similar situations'. This simple rule of brain function may set up behavioral trajectories that are difficult to change. A practical rule of thumb in this regard is that it takes about 1,000 repetitions of a new behavior before it becomes 'Automatic'. How many people deny themselves behavioral change by giving up too soon?

Currently, most models of brain dysfunction still focus on the spatial networks involved (for example, lesion–behavior relationships), or at the microscopic scale of the synapse and concomitant neurochemical imbalances.

Integrative Neuroscience is poised to develop and test more diverse models. Theoretical and numerical models of the whole brain, will increasingly allow evaluation of interrelationships among genetic, developmental, critical phases, learning, affect regulation, experience etc to be evaluated with respect to specific patterns of brain instability and symptom profiles. An integrative approach to this complexity, will allow content rather than fashion to elucidate the significant factors.

Biology and behavior both matter, and need to be understood. As more integrative dynamical models emerge and find experimental support, they will yield more precisely targeted treatments (be they pharmacological, psychotherapeutic or behavioral), that are based on increasingly fundamental principles.

## VI. CONCLUSION

The possibility of elucidating some of the integrative mechanisms of the brain as a dynamical system, no longer looks like such a pipe dream. But we will also be well served to be realistic about the rudimentary nature of current knowledge.

The cerebral 'black box' has now been well and truly opened. There are already a myriad advances in the basic neurosciences. Advances in physics and computing have made the objective measurement of overall brain structure and function accessible. In this era of Integrative Neuroscience, interpretation of such data will need to be coupled with a deep understanding of the brain's functions — which goes way beyond the common platitude that multiple regions in the brain are connected and interact.

In this Big Picture chapter, I highlighted the possibility of different rules operating and co-existing at different scales of function. In section II a spectrum of mechanisms across scale were outlined. There may of course be crucial mechanisms at every scale from quantum to macroscopic, that we still have no clue about — but it is hard not to be impressed by this detailed information explosion. Section IV showed an integration of the temporal and spatial information flow in whole brain models across disciplines. Needless to say many details remain unresolved and despite preliminary hints of integrative mechanisms, there remains a long way to go. Nevertheless, integrations within and across scale seem timely, and incorporation of the essence of such integrations into numerical brain models is now achievable. There is also a seminal variable that most conceptualizations have found too hard to incorporate, namely dynamics. The integrative *timing* of its network interactions is what constitutes coherent brain function.

The limitations of our understanding are no longer technological, nor due to the absence of ideas: they are conceptual. Existing ideas need to be brought together in more rigorous, quantitative and testable models. The central issue now is the capacity of testable models to link and generalize brain related information. We can anticipate Integrative Neuroscience contributing to increasing multidisciplinary participation among biological, psychological, ANN and numerical modeling efforts.

Advances in neuroscience across-scale are also being brought together into databases, making them more accessible (including on the world-wide-web) for inclusion into integrated models of the human brain. This emerging field of Neuroinformatics (Koslow and Huerta, 1997) is another reason to expect convergence within the brain sciences.

As participants from different disciplines become more familiar with each other's perspectives, models and data, achieving the goals of Integrative Neuroscience will be accelerated. I hope the perspectives and core information across disciplines provided in this book will contribute one small step in that direction.

## REFERENCES AND SUGGESTED FURTHER READING

Adams, RD, Victor M, Roper AH (1997) *Principles of Neurology*. McGraw Hill.

Amit DJ (1989) *Modeling Brain Function*. Cambridge University Press, New York.

Alexander GE, Crutcher MD and DeLong MR (1990) Basal ganglia-thalamocortical circuits: Parallel substrates for motor, oculomotor, 'prefrontal' and 'limbic' functions. *Progress in Brain Research,* **85**(6), 119.

Andreason, NC (1997) Linking mind and brain in the study of mental illnesses: A study for a scientific psychopathology. *Science,* **275**, 1586–1593.

Barinaga M (1998) Listening in on the brain. *Science,* **280,** April 376–378.

Barkley RA (1998) Attention Deficit Hyperactivity Disorder. *Scientific American* September, 44–49.

Basar E and Bullock TH (1992) *Induced rhythms in the brain.* Birkhauser, Boston.

Basar-Eroglu C, Struber D, Schurmann M, Basar E (1996) *International Journal of Psychophysiology,* **24,** 101–112.

Bliss TV and Lomo T (1973) Long-lasting potentiation of synaptic transmission in the dentate area of the anaesthetized rabbit following stimulation of the perforant path. *Journal of Physiology,* **232**(2), 331–56.

Boucsein W (1992) *Electrodermal Activity.* New York: Plenum Press.

Briggs J and Peat FD (1989) *Turbulent miror: An illustrated guide to Chaos Theory and science of wholeness.* Harper and low, New York

Carlsson M and Carlsson A (1990) Interactions between glutamatergic and monoaminergic systems within the basal ganglia — implications for schizophrenia and Parkinson's disease. *Trends in Neuroscience* (TINS), **13**(7), 272–276.

Carlsson A, Hansson LO, Waters N, Carlsson ML (1997) Neurotransmitter aberrations in schizophrenia: New Perspective and therapeutic implications. *Life Sciences,* **61** (2), 75–94. Elsevier Science Inc. USA.

Charney DS and Deutch A (1996) A functional neuroanatomy of anxiety and fear: Implications for the pathophysiology and treatment of anxiety disorders. *Clinical Reviews in Neurobiology,* **10**(3–4), 419–446.

Cloninger CR, Svrakic DM and Pryzybeck TR (1993) A Psychobiological Model of Temperament and Character. *Archives of General Psychiatry,* **50,** 975–990.

Cloninger CR (1994) Temperament and Personality. *Current Opinion in Neurobiology,* **4,** 266–273.

Churchland PS, and Sejnowski TJ (1994) *The Computational Brain.* MIT Press.

Damasio AR (1994) *Descartes' Error.* Picador.

Edelman G (1987) *Neural Darwinism. The theory of neuronal group selection.* Oxford University Press.

Eckhorn et al (1998) *Biological Cybernetics,* **60,** 121–130.

Franklin S (1998 ) *Artificial Minds.* The MIT Press, Cambridge, Massachusetts, USA.

Freeman WJ (1995) *Societies of Brains. A study in the neuroscience of love and hate.* Lawrence Erlbaum Associates, New Jersey.

Gardner H (1993) *Multiple Intelligences: The theory in practice.* Basic Books, New York

Gazzaniga MS (1994) *Nature's Mind.* Penguin Books.

George MS, Post RM, Ketter TA, Kimbrell TA and Speer AN (1998) Neural mechanisms of mood disorders. In: *Mood and Anxiety Disorders.* (Ed). Rush J. Williams and Wilkins Philadelphia, pg. 1–21.

Goldberg E (1990) *Contemporary neuropsychology and the legacy of Luria.* Lawrence Erlbaum Associates, Hillsdale New Jersey.

Goldberg E, Harner R, Lovell M, Podell K, Riggio S (1994) Cognitive bias, functional cortical geometry, and the frontal lobes: Laterality, sex, and handedness. *Journal of Cognitive Neuroscience,* **6**(3), 276–296.

Goleman D (1996) *Emotional Intellegence: Why it can matter more than IQ.* Bloomsbury.

Gordon E and Rennie C (1999) *Integrative Neuroscience and Purposeful Dynamical Networks* (submitted).

Gray JA (1995) The contents of consciousness: A neuropsychological conjecture. *Behavioral and Brain Sciences,* **18,** 659–722.

Gottfredson LS (1998) The General Intelligence factor. *Scientific American,* **9**(4), 24–29.

Gray JA (1998) Integrating Schizophrenia. *Schizophrenia Bulletin,* **24**(2), 249–266.

Gray JA (1982) *The Neuropsychology of Anxiety: An Enquiry into the Functions of the Septohippocampal System.* Oxford University Press.

Halgren E and Marinkovic K (1994) Neurophysiological Networks Integrating Human Emotions. In: *Cognitive Neurosciences* (Ed.) Gazzaniga MS.

Hebb D (1949) *The Organization of Behavior: A neuropsychological Theory.* Wiley, New York.

Kandel ER, Schwartz JH and Hessell TM (1991) *Principles of neural science.* Prentice-Hall: London.

Kandel ER and Hawkins RD (1992) The biological basis of learning and individuality. *Scientific American,* **267**(3), 52–61.

Kauffman S (1993) *The Origins of Order.* Oxford University Press, New York.

Koslow SH and Huerta MF (1997) *Neuroinformatics; An overview of the Human Brain Project.* Lawrence Erlbaum Associates.

Langton C (1990) Computation at the Edge of Chaos: Phase transitions and emergent computation. *Physica D,* **42**, 12–37.

LeDoux JE (1997) Emotion, Memory and the Brain. *Scientific American,* Mysteries of the Mind; June, 68–75.

LeDoux JE (1998) *The Emotional Brain: The Mysterious Underpinnings of Emotional Life*: Touchstone.

Luria AR (1973) *The Working Brain.* Penguin Books.

Maturana H and Varela F (1987) *The Tree of Knowledge.* Shambhala, Boston.

Mayberg HS (1997) Limbic — cortical dysregulation: A proposed model of depression. *Journal of Neuropsychiatry and Clinical Neurosciences,* **9**(3), 471–481.

Ohman A (1979) The orienting response, attention, and learning: An information-processing perspective. In *The Orienting Reflex in Humans* Ed. Kimmel HD, Van Olst EH, Orlebeke JF. Lawrence Erlbaum Associates, New Jersey.

Phillips ML, Young AW, Senior C, Brammer M, Andrew C, Calder A, Bullmore E, Perrett D, Williams S, Gray J and David A (1997) A Specific Neural Substrate for Perceiving Facial Expressions of Disgust. *Nature,* **389**, 495–498.

Pliszka SR, McCracken JT and Maas JW (1996) Catecholamines in Attention Deficit Hyperactivity Disorder: Current Perspectives. *Journal of the American Academy of Child and Adolescent Psychiatry,* **35**(3), 264–272.

Porges SW (1998) Love and the Evolution of the Autonomic Nervous System: The Polyvagal Theory of Intimacy. *Psychoneuroendocrinology ,* **23**, 837–861.

Posner MI and Petersen SE (1990) The attention system of the human brain. *Annu. Rev. Neuroscience,* **13**, 25–42.

Prigogine I (1989) *From being to becoming: Time and complexity in the physical sciences.* San Fransisco.

Schore A (1994) *Affect Regulation and the Origin of Self.* Lawrence Erlbaun Associates, New Jersey.

Seligman MEP (1994) *What you can change and what you can't.* Random House.

Shepard GM *et al.* (1997) Senselab: A project in multidisciplinary multilevel sensory integration. In: *Neuroinformatics: An overview of the Human Brain Project.* Lawrence Erlbaum Associates, New Jersey.

Shiffrin RM, Schneider W (1977) Controlled and automatic human information processing: Perceptual learning, automatic attending and a general theory. *Psychology Review,* **84**, 127–190.

Singer W (1995) Development and plasticity of cortical processing architectures. *Science,* **270**(3), 758–775.

Sokolov EN (1960) Neuronal models and the orienting reflex. In: *The Central Nervous System and Behavior.* Brazier, MAB, New York, Josiah Macy Jr. Foundation:187–276.

Sokolov EN (1963) *Perception and the conditioned reflex.* New York, MacMillan Company.

Sokolov EN (1975) The neuronal mechanisms of the orienting reflex. In: Vinogradova OS, ed. *Neuronal Mechanisms of the Orienting Reflex,* Hillsdale NJ: Lawrence Erlbaum: pp 217–235.

Sokolov EN (1990) The orienting response and future directions of its development. *Pavlov J Biol Sci,* **25**, 142–150.

Sperry RW (1950) Neural basis of the spontaneous optokinetic response. *Journal of Comparative Physiology,* **43**, 482–489.

Von Bertalanffy L (1968) *General Systems Theory.* Braziller, New York.

Wiener N (1948) *Cybernetics.* MIT Press.

# Chapter 2

# OVERVIEW OF CHAPTERS

Evian Gordon

The Brain Dynamics Centre, Westmead Hospital and
Department of Psychological Medicine, The University of Sydney

This book represents a survey of brain models covering the full range of specializations, and points in the direction with which they might begin to be brought together (in a preliminary unified manner), tested and further developed.

The book includes models from basic sciences and from applied clinical disciplines. The initial chapters deal with the basic sciences, physical anthropology, philosophy, anatomy, chemistry and physiology. Computer models and brain dynamics chapters then examine how these parts contribute to overall adaptive function. They are followed by chapters describing more applied clinical models that have arisen in psychology, neurology and psychiatry. It will be seen that, although the models are diverse in nature, as are the data against which they are tested, they all represent valid attempts to organize our knowledge of the brain.

The chapters are on:

3)      Evolution of the Human Brain.
4)      The Mind–Brain Problem.
5)      A Cellular Perspective of Neural Networks.
6)      The Brain's Chemistry.
7)      The Brain's Anatomy.
8)      Sensory-Motor Models of the Brain.
9)      Computer Models of the Brain.
10)     Brain Dynamics.
10a)    Modeling the Whole Brain in Action.
10b)    Brain Chaos and Intentionality.
11)     Models of the Brain in Neurology.
12)     Models of the Brain in Psychology.
13)     Models of the Brain in Psychiatry.
14)     Mind–Brain in Psychotherapy.
15)     Human Brain Imaging Technologies.

In chapter three (Evolution of the Human Brain), Phillip Tobias outlines the astonishing trebling in hominid brain size (and associated changes in macroscopic and microscopic brain organization) that have occurred during the relatively brief evolutionary period of the last 3 million years; from Australopithecus to Homo habilus to Homo erectus to Homo sapiens. No other species has undergone changes in the brain of this magnitude, over the past few million years (the dolphin for example reached its current stage of brain development some 20 million years ago). Standing upright and the freeing of our hands with Australopithecus (and the major structural and functional changes that must have resulted), would be consonant with the importance of placing sensory-motor processing at the core of the brain's functions (as suggested in chapters 8 and 10b). The dramatic increase in brain size began after the advance of bipedalism. The sheer magnitude of the brain's changes should compel us to be mindful of the impact that evolution has had in shaping the human dynamical brain. This guiding perspective in physical anthropology, should be coupled with possible clues from social anthropology (for example the natural selection of increased inhibitory brain functions to facilitate group cohesion), and provide a context for all modeling efforts.

In the fourth chapter on The Mind–Brain Problem, Peter Slezak explores the history of this conundrum by contrasting the two main approaches to this problem, Dualism and Materialism. Dualism states that the mind is unlike the body, perhaps a spiritual substance, whereas with Materialism, the mind must be explained in terms of the physical workings of the material brain. Slezak focuses on Cartesian Dualism, with variants such as Epiphenomenalism and Parallelism or other approaches to the problem, such as Idealism, beyond the scope of this chapter. The 'ghost' in the dualist machine was initially exorcised by the Behaviorists, by excluding the brain from consideration entirely and examining only behaviors in relation to stimulus–response relationships. This approach was superseded by Materialism, beginning with Identity Theory, which stated that mental events were identical with brain events. This ruled out in principle the possibility that anything else, besides brains, might have a mind. This led to the Functionalist Theory, that the mind is not to be identified directly with the brain, but rather with the abstract features of brain function (i.e., software or process). A significant challenge to this Functionalist symbolic version of cognition, has taken the form of 'Connectionist' models, which are more faithful to the interconnected biological workings of the brain. The spirit of this book is to explore the integrated interrelationships of the human brain ('content') and the mind ('form, interconnected pattern and process').

The history of brain science has primarily been shaped by the information explosion at the microscopic single neuron and molecular scale. This data remains the solid foundation for all of brain science. This book focuses on what happens when 10 billion of these neurons are brought together into a single system. In chapters five and six, Glenn Hunt and Philip Beart explore the core building blocks of neuroscience, in A Cellular Perspective of Neural Networks and The Brain's Chemistry.

Most neuroscientists focus exclusively on the single neuron microscopic scale — probing the electrical activity, ion channels, brain chemistry and metabolic pathways associated with neural activity. Others record and model the electrical transmembrane potentials and their propagation through the dendritic branches to the cell body, where action potentials are triggered (only when the transmembrane gradient is lowered to a critical threshold of minus 50 millivolts), and then travel at constant speed down the axon to its end branches, where it precipitates release of stored brain chemicals (neurotransmitters), that traverse a tiny gap (synapse) to form the communication link to connected neurons. When these neurotransmitters arrive at the connected neurons' receptors, they begin a cascade of effector, second and third messenger processes all the way to the cell's genes. All information transfer in the brain consists of these electrochemical processes. Since the speed and size of the action potentials are constant, a critical communication code between neurons, is the rate of firing of action potentials. The consequent effects of the released neurotransmitters are either excitatory, inhibitory or modulatory and the ongoing excitatory–inhibitory balance in networks underlies all brain functions.

All these processes exist at the multiplicity of co-existing scales of organization in the brain. It is easy to lose sight of the fact that every fraction of a second, each single neuron in the brain is making a decision about whether or not to fire. These outcomes reflect ongoing feedforward and feedback processes at the whole brain scale, continuously adapting to the environment. Each scale of function might involve different mechanisms and therefore demand different approaches to modeling. In chapter 10b we shall see that the function of a single neuron is highly non-linear, but feedforward and feedback processes at the neural network and whole brain scale result in a rich cast of linear as well as non-linear processes. Integrative Neuroscience will grapple with the manner in which this multiplicity of mechanisms co-exist, but the parameters associated with individual neurons (time course of action and signaling processes) and their gene functions, will remain fundamental building blocks in modeling the brain.

In the seventh chapter, The Brain's Anatomy, Ken Ashwell, Liz Tancred and George Paxinos present a model of the organizational plan of the brain. The chapter outlines the core architecture. The brain is shown essentially to consist of Sensory Input–Motor Output with Association networks in-between. Sensory input networks convey information from the environment via relay nuclei in the thalamus (and projection networks in the brainstem) to the specialized sensory networks in the back portions of the cortex. These processed signals are transmitted to the front portions of the brain for motor output (undertaken in conjunction with the basal ganglia and cerebellum). Output also includes speech and autonomic output to control the metabolic and hormonal functions of the body. In between these Input and Output networks are the vast majority of the human brain's neurons, the Association networks (comprising the prefrontal, parieto–occipito–temporal cortex and limbic system association cortex).

Whilst this simplified model will be shown to be useful, it fails to reveal a key fact about the brain — that it is a very highly (mainly bi-directional) interconnected system, both in terms of its anatomy and chemistry. The precise details of these interconnections are extremely relevant as specific clues in realistic models of the human brain. The details of some of the more unidirectional flows are also essential to our understanding of the brain. These are found for example in the motor pathways, where direct and indirect motor pathways link the cerebral cortex to the striatum, then to the thalamus and back to cerebral cortex (see Alexander model in the Big Picture); monoaminergic projections are predominantly unidirectional, ascending from serotonergic, dopaminergic and noradrenergic nuclei in the brainstem to distribute widely throughout the brain. The limbic system engages in connections with the cortex indirectly via the parahippocampal gyrus and subiculum, with outputs from its loop also going to the hypothalamus and brainstem. The amygdala engages in direct reciprocal connections with the frontal and temporal lobes. The hippocampus has input from neocortex, but mainly indirectly through the anterior parahippocampal gyrus, which receives many association fibres from the medullary core of the hemisphere. Similarly, output to isocortex from the hippocampus is very indirect, being channeled through the thalamus and cingulate gyrus.

Such structural details are essential to brain models and offer perspective about bi-directional interconnectivity, as well as hints about the functional organization of the brain and its overall flows of activity.

In chapter eight (Sensory-Motor Models), Alan Freeman outlines how the brain can be regarded as being built around a core of sensory-motor networks. Sensory-motor interactions consist of a hierarchy of reflex loops, upon which more sophisticated functions occur. The sensory networks filter relevant information from the environment at the receptor level, this data is relayed via the thalamus and brainstem to the sensory cortex — where signal streams are combined in the association cortex and compared with stored data from previous experience to provide object or sound recognition and localization. Motor output requires transformation of an overall plan of action into contraction of individual muscles. A plan is initiated in the supplementary motor area, premotor cortex and basal ganglia. These networks engage the primary motor cortex for fine control of individual muscles via the corticospinal tract, and simultaneously engage the cerebellum for adjustments to overall posture. These processes are however not undertaken in a serial manner. Feedforward and feedback processes in the sensory-motor networks operate in a parallel fashion, to allow coordination and error correction in our ongoing adaptation with the environment.

The ninth chapter outlines Computer Models of the Brain. William Schmidt and Evian Gordon show how the advent of computer power has spawned numerous computer simulation approaches to model the brain. The 'Symbolic manipulation' approach is extensively used in the fields of cognitive psychology and artificial intelligence (AI) and is also consistent with the Functionalist approach (in the Mind–Brain chapter). It proposes that the logical operations performed by the brain can be simulated on computers, with the human brain being seen as one of a range of possible realizations of an intelligent system and so specific biological information about the brain is downplayed. Cognition is seen as a high level phenomenon (software rather than hardware), that can be treated independently of any particular implementation details.

In 'Connectionism' on the other hand, real neural networks owe their effectiveness to the sheer number of interconnected neurons that they contain and to the adaptable strengths of their synapses. There is a range in the extent to which biological information is used in Connectionist approaches. In the Artificial Neural Network (ANN) approach very simplified neurons are used and surprisingly powerful image recognition and learning algorithms have been developed. In the emerging field of 'Computational Neuroscience' on the other hand, models use more detailed biological information. It is within Connectionist approaches that an increasing number of scientists are predicting an integration of AI and biological approaches. Patricia Churchland and Terence Sejnowski (1994) state that 'If we are to understand how the brain sees, learns, and is aware, we must understand the architecture of the brain itself. The computational style of the brain and the principles governing its function are not manifest to a casual inspection. Nor can they be just inferred from behavior, detailed though the behavioral descriptions may be, for the behavior is compatible with a huge number of very different computational hypotheses, only one of which may be true of the brain. Moreover, trying to guess the governing principles by drawing on existing engineering ideas has resulted in surprisingly little progress in understanding the brain, and the unavoidable conclusion is that there is no substitute for conjuring the ideas in the context of observations about real nervous systems: from the properties of neurons and the way neurons are interconnected'.

The chapter offers an historical perspective of ANN computer models in Strong AI and Weak AI. It also outlines more biologically faithful approaches to computer models in Connectionism and Computational Neuroscience (including Neuronal, Network and Whole Brain models). The chapter also provides a platform from which to integrate emerging current ANN models of network dynamics (Amit, 1998), whole brain ANN models (Franklin, 1998) and biological models of the brain as a dynamical system (outlined in the next two chapters).

The critical and often neglected aspect of the brain in modeling efforts, is the dimension of *time*. The brain is a highly evolved dynamical adaptive system. It seems unlikely that the brain evolved to perform primarily logical operations. Rather, it seems to have evolved because its dynamical properties and structural connections to and from the environment conspired to keep it out of trouble and reproduce its own kind.

In chapters 10a and 10b, Jim Wright and Walter Freeman point the way forward for the emerging field of 'Brain Dynamics'. They present models and perspectives as to how the brain operates as a dynamical system. These models use realistic details of brain anatomy and physiology. They emphasize the dynamic laws of interaction between neurons over a time scale of a fraction of a second — which is the time scale at which cognition actually occurs. These models have a special respectability that comes from being numerical and being tested against real brain data. Wright's model focuses on linear dynamics, local-global interactions and modulations by reticulo-cortical feedback loops. Freeman's model emphasizes non-linear dynamics and includes the setting of goals by the limbic system and their implementation by cortical networks.

These models also demonstrate an impressive bridging of scales, highlighting that the first broad outline of the dynamic interactions of billions of neurons is now discernible. They constitute preliminary exemplar 'engines' of a working dynamical brain. Such models can be expanded (parameters added) by researchers in different disciplines. Interdisciplinary researchers could readily add the most relevant elaborations of anatomical detail, neurotransmitter receptor type and specific feedbacks in the case of the Wright, Robinson, Rennie *et al.* model (which is numerical — so direct comparisons can be made with real data). Freeman also proposes specific directions for extending his model.

With the benefit of these core basic science models, we are ready to traverse the models of the human brain in the applied clinical disciplines of neurology, psychology and psychiatry.

In the eleventh chapter, Models of the Brain in Neurology, Victor Fung and John Morris undertake a journey through the brain exploring a Localizationist perspective. It is mainly since Broca's 1861 description of localized speech areas in the left hemisphere, that a consistent view has been propounded, that brain function could be understood as hierarchical serial assemblies of 'centers' which have a consistent anatomical localization. There do indeed seem to be specialized brain networks, particularly associated with the primary sensory-motor cortex and aspects of language. But Localizationist models unintentionally encourage the assumption that if function A is impaired due to injury of brain structure B, then B is the site of A in the normal brain. If that assumption is made, it is the fault of the assumer, not the model. The authors show that Localizationist models remain useful in some aspects of clinical practice. But they also highlight that many clinical disorders (for example a patient with Parkinson's Disease who could barely walk — but could dance) demand more holistic distributed brain models, which focus on the overall connectivity and interactions amongst the brain's networks.

In the twelfth chapter (Models of the Brain in Psychology) Iain McGregor and William Schmidt outline how psychology consists of a number of sub-disciplines that have borrowed, adapted and developed an array of models, incorporating Localizationist, Distributed, Symbolic and Connectionist approaches. However, the distinguishing focus of these models is on high-level brain functions and their possible underlying mechanisms. The phenomena of interest include perception, attention, lateralization of function, information processing, learning, memory, language and emotions. Selected mainstream models of these high-level phenomena are summarized.

On the broadest perspective, Models of the Brain in Psychiatry (described by Graeme Smith in chapter thirteen) are couched within a Biopsychosocial Systems Theory. Nevertheless, the strength of this eclecticism is also its weakness, since few explicit hypotheses that traverse biological, psychological and social domains have found experimental support. Over the past two decades there has been heightened interest in models that propose a biological basis for psychiatric disorders. They essentially focus on imbalances in brain chemistry underlying disorders such as anxiety, depression and schizophrenia. Anxiety disorders are associated with increased activation of the 'fight or flight' reflex, norepinephrine and involvement of GABA. Depression is associated with decreased norepinephrine and serotonin and imbalances in dopaminergic movement networks. Schizophrenia is associated with a failure in integrated processing, increased dopamine, imbalances between glutamate and dopamine, and disturbed modulation of serotonin. Model development and testing (included with brain imaging technologies) is embryonic, but likely to escalate rapidly with the influx of multidisciplinary personnel and more integrated models into this field.

Russell Meares describes Mind–Brain in Psychotherapy in chapter fourteen. He presents three broad models. Freud's model (and its subsequent adaptations) are 'Ego Models' which focus on reducing the tension between intrapsychic instincts, with repression acting as a barrier between conscious and unconscious processes. Klein and

Fairburn's 'Object Relations Model' considers recurring disturbed patterns of relationships with others as a central tenet. 'Models of Self' (generated by James, Jung, Winnicott and Kohut) see development of a cohesive sense of Self as fundamental to normal psychological development. The high level of abstraction in models of psychotherapy has allowed proponents in this field to protect some of their more cherished beliefs from scientific scrutiny. The emergence of a multidisciplinary cognitive science, provides a landscape for models of mind and brain to be systematically explored and tested. Some of these models are testable, albeit with difficulty. In Ego models, aspects of the subconscious might be explored via 'Automatic' processing. 'Reattributions' of negative thoughts to more appropriate ones in Cognitive Behavior Therapy (associated with Object Relations model) are readily quantifiable, and Russell Meares provides a compelling case for the use of language as one means of testing models of Self. In terms of brain dynamics, psychotherapy focuses on interpersonal interactions. Integrative Neuroscience might help to provide cogent distillations of such information, to diminish the current gap between metaphor and brain mechanisms in this field. As this gap is narrowed, the prospect of an 'Integrative Psychotherapy' among its diverse branches, that is tailored to each individual's situation, will become more likely.

In the final chapter Evian Gordon, Chris Rennie, Arthur Toga and John Mazziotta outline the range of brain imaging and recording technologies that allow testing of many human brain models at the whole brain scale. These technologies (CT; MRI; fMRI; EEG; MEG; RCBF; SPECT; PET) allow measurement of brain anatomy, electrical brain function and radioisotope based measurements of brain chemistry and drug activity. The complementarity of these functions and technologies is highlighted.

One exemplar International Database is described (The International Consortium for Brain Mapping). A number of inevitable directions are outlined, that seem likely to enhance the usefulness of these techniques, and cautions about simplistic non-model-driven interpretation of data are presented.

# Chapter 3

# EVOLUTION OF THE HUMAN BRAIN

Phillip V Tobias

Palaeo-anthropology Research Unit, Department of Anatomy and Human Biology
University of Witwatersrand Medical School

The brain qualifies as one of Charles Darwin's 'organs of extreme perfection and complication'.

The problem of how structure–function composites of astonishing complexity could have evolved gradually and by natural selection has been with us at least since Darwin admitted how difficult it was to explain, on 'his' theory, the origin of the eye of higher animals. The human language capacity is another evolutionary achievement of extraordinary perfection and complexity. Like other human skilled activities, it involves both central (neural) and peripheral (vocal and respiratory) complexes. The reduction of these to simpler building-stones to which evolutionary principles may be applied is staggeringly difficult. It is a much more formidable task to envisage and reconstruct the evolution of the brain than, say, that of the bones and the teeth. Under suitable conditions of rock and soil, the latter organs undergo fossilization and the study of well-dated fossil remains permits us to piece together the story of how bones and teeth have changed through vast aeons of time. Brains, however, do not fossilize. Where then do we find the evidence from which to infer how brains have evolved?

One source is provided by the comparative anatomy of the brains of living species including *Homo sapiens*. For example, modern human brains are, in absolute size, about three times as large as those of the living great apes. From such data it is reasonable to infer that, if we and the great apes have in the past had common ancestors, the curious features of the human brain must have emerged after the lineage leading to *Homo sapiens* diverged from the lines of descent of the apes.

The fossil record enables us to confirm inferences about the structure and function of the brains of ancient members of the human lineage. The curious and unique relationship between the brain and the calvaria (skull) is the key to such studies.

## BRAIN AND BRAIN-CASE

The anatomist reads and interprets the markings and impressions of ligamentous and muscular attachments to bones, as well as the grooves, notches and smoothings owing to the impingement on bones of other structures such as arteries, veins, nerves and tendons in transit. In the same way, much information on functionally important soft tissue anatomy may be garnered from the careful study of even the most ancient fossilized bones.

Perhaps of no structure has this proved more valid than the brain, because in the skull we have a bony complex that in life is faithfully moulded upon the contents, namely the brain and its coverings, blood vessels (vasculature) and blood, cerebrospinal fluid and the stumps of emerging cranial (brain) nerves. Under conditions of normal development, a larger brain dictates the growth of a larger skull and a smaller brain is housed in a smaller skull. A broader brain is accommodated in a broader brain-box. A subject whose brain has a relatively larger region (such as the frontal lobe) has a proportionately larger skull in that area. Brain structures stamp telltale grooves on the skull.

All such features of the soft tissue contents of the skull may readily be confirmed in modern cadavera. Hence, from the interior of a skull, one can draw many conclusions about the brain and vessels that once occupied that brain-case. We can facilitate the process by filling an empty brain-case with a plastic or plaster medium and making an artificial endocranial cast. In size and form such an endocast faithfully reproduces the size and form of the endocranial cavity. The surface of the endocast takes an impression of all the markings that have been imprinted during life on the endocranial surface. Hence one may read the brain's convolutions (sulci and gyri) and vascular impressions directly from the surface of the endocast.

Sometimes a natural endocast forms during fossilization. Once the soft tissue contents of the brain-case have disappeared, sand may gain access to the calvaria. If chemicals such as lime are present in the surrounding rocks, the sandy filling of the braincase becomes calcified. This results in a natural endocast. Details of the brain and vessels may be read from a natural endocast, as from an artificial endocast. Endocranial surfaces and endocasts provide raw data for palaeoneurobiological research.

## THE ENDOCRANIAL CAPACITY

The most obvious results of endocranial studies are the endocranial capacities of hominoids. (Hominoids or the Hominoidea is a taxonomic grouping which comprises existing and ancestral apes and humans.) One speaks of capacities and not brain sizes because the brain-case includes other tissues besides the brain. Where data for actual brain weight or volume are available, there is a very high correlation between these values and the capacities. Because it is difficult to measure brain size and because we have thousands of dried crania of recent humans and apes, far more studies have been devoted to endocranial capacities than to brain volume or weight.

The author's estimates of the average values for endocranial capacities of combined-sex samples of modern hominoids (African apes, Asian apes, humans) are given in table 3.1 (Tobias, 1968, 1970, 1971). For male-plus-female samples, the modern human average capacity is 2.68 times the gorilla mean, 3.33 times the orang-utan value and 3.52 times the chimpanzee average.

TABLE 3.1.
Mean Endocranial Capacities of Hominoids, Past and Present.

| | Species: | Mean Volume (cm³) |
|---|---|---|
| **MODERN APES** | Chimpanzee (*Pan troglodytes*) | 383 |
| | Gorilla (*Gorilla gorilla*) | 504 |
| | Orung-utan (*Pongo pygmaeus*) | 404 |
| **FOSSIL APEMEN** | *Australopithecus afarensis* | 413 |
| | A. africanus | 440 |
| | A. robustus | 530 |
| | A. boisei | 463 |
| **EARLY HUMANS** | *Homo habilis* | 640 |
| | *H. erectus* | 937 |
| **MODERN HUMANS** | *Homo sapiens sapiens* | 1350 |

## ENDOCRANIAL CAPACITIES OF EARLY FOSSIL HOMINIDS

There are four widely recognized species of early apemen, namely *Australopithecus africanus* (whose remains from southern Africa are dated to approximately 3.0 to 2.4 million years ago [mya]); *A. afarensis* (4.0 to 2.8 mya); *A. robustus* (formerly called *Paranthropus robustus*, 1.8 to 1.5 mya); and *A. boisei* (formerly *Zinjanthropus boisei*, 2.5 to 1.3 mya).

From the more complete calvariae and natural endocasts of these four species of the australopithecines, it has been possible to arrive at values for their endocranial capacities, with varying degrees of reliability (table 3.1). The sample sizes for these early hominids are very small, compared with those available for living hominoids, so it is likely that our fossil series have sampled only a small part of the total variability within each species. Also, estimates of intraspecific sexual dimorphism will of necessity be very tentative.

These latest mean values (Tobias, 1994) confirm that the four australopithecine species show little real advance in absolute terms on the mean values for the existing great apes.

Thus, the *A. afarensis*, *A. africanus* and *A. boisei* mean values are, respectively, 1.08, 1.15 and 1.21 times the mean for the common chimpanzee; 0.82, 0.87 and 0.92 of that for gorilla; and 1.02, 1.09 and 1.14 times that for orang-utan. In sharp contrast, the modern human, interracial, male-plus-female mean is 3.52 times the chimpanzee mean, 2.68 times the gorilla value and 3.33 times the orang-utan average.

In other words, the earliest fossil specimens that, on dental, cranial and postcranial evidence, have been assigned to the Hominidae, had mean absolute capacities of the same order of magnitude as those of living great apes, but showed a small advance on the chimpanzee mean value. The mean capacity value for *A. afarensis* is about 8 per cent greater than that of modern chimpanzee, while the *A. africanus* mean is some 15 per cent greater than the chimpanzee mean. The chimpanzee is most closely related genetically to modern *H. sapiens*, while the gorilla is somewhat less similar. This is interpreted as indicating that the human lineage departed from that of the chimpanzee somewhat more recently than the divergence of the gorilla. The dating of the chimpanzee-human divergence is provisionally set between 6.4 and 5.0 mya, while the divergence of the gorilla lineage from the other African hominoids is dated somewhat earlier, between 9.0 and 6.0 mya.

By the time *A. afarensis* and *A. africanus* had emerged as upright-walking creatures with teeth of hominid pattern, their brain size had increased very slightly over that of the modern chimpanzee. These figures underline anew that *Australopithecus* comprised small-brained hominids and they refute Keith's (1915) expectation that dramatic brain enlargement would have been a hallmark of the earliest recognizable hominids. The dramatic increase in brain-size was to begin later, with the emergence of *Homo habilis* (Leakey *et al.*, 1964).

The latest estimates of the mean capacity of *H. habilis* show that this species had a brain size bigger by half than the average values in *Australopithecus*. Thus, it was with *H. habilis* that the human trend toward great cerebral expansion began. For *H. erectus* the mean value of 937.2 cm$^3$ is 46.4% greater than the sample mean for *H. habilis* (Tobias, 1994), whilst the estimated capacity in modern *H. sapiens*, irrespective of sex and race, namely 1350cm$^3$, exceeds the *H. erectus* sample mean by some 44%.

It would be wrong to conclude from these data that the increase in brain size over

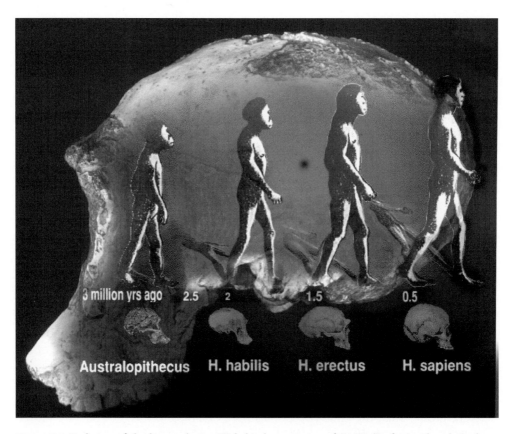

**Figure 3.1** Evolution of the human brain. With kind permission of PAGE (Professional and Graduate Education, University of Wollongong).

the last three million years occurred by a series of quantum leaps with periods of stasis or equilibrium between the supposed waystations of hominid encephalization. Each palaeo-species lived for a time span and some specimens in a sample have been derived from earlier stages, and others from later stages, in the time of that species on the planet.

Among the hominid species, there is considerable overlap in time: furthermore, most scholars agree that *A. robustus* and *A. boisei* (which some would place in a separate genus *Paranthropus*) have specialized away from the lineage leading to *H. sapiens*. Although they are hominids, they are not direct ancestors of modern humans. There is appreciable overlap in the ranges of endocranial capacity values and in the 95% population limits. Hence, the simple sequence of mean values: *A. afarensis–A. africanus–A. robustus–A. boisei–H. habilis–H. erectus–H. sapiens* gives an over-simplified picture of the pattern of change. However, the general hominid trend towards higher capacity with time is strikingly evident.

Within any one species, no convincing trend appears to be displayed from the earlier to the later specimens assigned to that species, but sample sizes in these sub-sets are very small. Only when larger samples for each species become available will it be worthwhile and justified to see if intraspecific trends can be detected.

## RELATIVE BRAIN SIZE

Is the difference in capacity, between the very early hominids and existing great apes, 'real', or is it simply the consequence of differing body sizes?

A number of different techniques have been proposed to determine the degree of encephalization (brain size, when body size is taken into account). Lashley (1949) had suggested that the total amount of brain material, expressed as a fraction of total body size, 'seems to represent the amount of brain tissue in excess of that required for transmitting impulses to and from the integrative centers'. Based on this concept, Jerison (1963) proposed a formula for the calculation of the 'excess' neurons. First, it was necessary to know the total brain size and the total body size for a number of species in a major systematic group. From these data, Jerison attempted to estimate, for a single species in that group, the number of neurons associated with improved adaptive capacities.

When we deal with fossil taxa, body size (sometimes expressed as stature, more commonly as body mass or weight) is generally calculated from the dimensions of postcranial bones. This is more easily said than done and some of the difficulties have been discussed elsewhere (Tobias, 1994).

An immense literature has been devoted recently to 'scaling', that is, the structural and functional consequences of differences in size (or scale) among organisms of more or less similar design (Jungers, 1985). Lande (1985) observes that 'genetic uncoupling' of brain and body sizes in primates would have facilitated encephalization in primates, because natural selection for larger brain size would then not necessarily have carried along an uneconomical, correlated increase in body size: ' … if the genetic correlation between brain and body size within populations in the human lineage was … low as suggested by the data on primates, hominids would have been enabled to rapidly increase brain size in response to selection for more complex behavior without the cost of antagonistic selection to prevent the evolution of gigantism' (Lande, 1985: pp. 30).

Estimates of relative brain size show that (a) the various australopithecine species were slightly more encephalized than the chimpanzee; and (b) *H. habilis* was clearly more encephalized than any of the australopithecine series and represented a major step, indeed the first such, in brain aggrandizement: its values reveal that it had attained some 50 % of the *H. sapiens* degree of encephalization (Tobias 1987). More marked encephalization followed from *H. habilis* to *H. erectus*, the latter species reaching some 70% to 80% of the degree of encephalization shown by *H. sapiens*.

*H. habilis* is thus appreciably advanced in its degree of encephalization as compared with *A. afarensis* and *A. africanus*. Since the estimated body size is built into the encephalization formulae, the larger endocranial capacity of *H. habilis* is not solely the result of its larger estimated body mass nor of a higher estimated stature (Grüsser and Weiss, 1985): it clearly represents a real advance in encephalization over the small-brained australopithecines.

The data on relative brain size show that, while the australopithecines were encephalised slightly more than the chimpanzee, *H. habilis* had unequivocally begun the remarkably 'uncoupled' or disproportionate enlargement of the brain that is a critical hallmark of humankind.

## THE ANATOMY OF HOMINISATION

There is good evidence that the early hominid species (*A. afarensis* and *A. africanus*) had developed adaptations to erect, bipedal stance and gait, though with a persistence of upper limb traits suggesting some residual overhead supportive and perhaps tree-climbing activities (Berger, 1994). Likewise, the evidence of the lower part of the vertebral column and its articulation with the pelvis shows that the weight-bearing and weight transmission functions of *A. africanus* were incompletely adapted to the erect posture and bipedal gait (Benade, 1990). In these features, the australopithecines were twilight creatures, hominid in their bipedalism and their teeth, ape-like in their arboreal vestiges and small brains. It is therefore pertinent to enquire whether their brains showed signs of hominid structural features.

The increase of brain size, whether absolute or relative to body size, is certainly the most dramatic change to have occurred in hominid evolution in the last three million years. In that period, spanning some 200,000 generations, brain size trebled along the human lineage.

The advantages of the larger size have been much speculated upon and over a dozen hypotheses have been advanced to explain the sustained tendency in our lineage towards increased encephalization. However, an encephalizing trend connotes not only that a bigger percentage of one's bodily bulk is occupied by brain tissue. For size change is only a gross, external indicator of encephalization. In modern animal groups, when comparisons are made between smaller and larger brained species, it is found that, with larger brain size, there occur an increase in the number of neurons, the dendrification and connectivity, the glia–neuron ratio, and with the latter, a decreased packing density of the neurons. It is reasonable to infer that changes in these features must have occurred during hominid encephalization. In addition, direct observations on fossil endocasts tell us what reorganization of the surface of the brain, mainly of the cerebrum, has occurred during encephalization.

## FUNCTIONAL AREAS OF THE AUSTRALOPITHECINE BRAINS

Five principal morphological characters distinguish australopithecine endocasts from the brains of existing apes (see chapter 7 for orientation to the brain's basic architecture):

(1)     The brainstem is situated further forward than in apes.

(2)     The parietal lobe of the cerebrum is well-developed (Holloway, 1988).

(3)     The cerebellar hemispheres are underslung (Tobias, 1967) so that the occipital poles of the cerebrum almost always form the most backward part of the endocast.

(4)     Show right fronto-petalia (that is, the frontal pole of the right hemisphere protrudes further forward than that of the left hemisphere) and left occipito-petalia (in which the occipital pole of the left hemisphere protrudes further backward than that of the right hemisphere). This combination Galaburda (1984) describes as the most common in modern humans, whilst Holloway (1988) declares it is not found in the apes.

We cannot be sure what cyto-architectonic (neuronal detail) differences underlay this fourfold restructuring of the brain of various australopithecines. Suffice it to say that these superficial rearrangements, with a rather small increase in absolute and relative endocranial capacity, are the total available information on differences between the brains of australopithecines and of present-day chimpanzees.

## THE BRAIN OF *HOMO HABILIS*

*H. habilis* not only possessed an appreciably enlarged brain as compared with *A. africanus,* but also showed numerous morphological changes of the endocasts that point to major restructuring of the brain in a modern human direction, including:

(1)     A broadening (mainly of the frontal and parietal lobes of the cerebrum), and a moderate heightening, but scarcely any lengthening of the cerebral hemispheres.

(2)     The sulcal pattern of the frontal lobes is similar to that of modern *H. sapiens* and quite different from that of existing apes.

(3)     The gyral impressions on the frontal lobe include a well-marked prominence in the posterior part of the inferior frontal convolution, in the position of Broca's area (a speech area).

(4)     There is a right fronto-petalia in the few *H. habilis* endocasts in which left and right frontal poles are preserved.

(5)     The impression of the parietal lobule (associated with spatial orientation) is well developed (Tobias, 1987).

(6)     The impressions of the supramarginal and angular gyri, comprising the inferior parietal lobule, are present for the first time in the hominid lineages. Since this area forms part of the larger Wernicke's area (a speech area in the cortex), it has been claimed that — with the speech cortex of Broca present as well — *H. habilis* is the first species in the history of the hominids to show the two most important neural bases for language abilities (Tobias, 1987, 1991).

The most important morphological traits of the *H. habilis* brain are the presence of the two main cerebral areas that in modern humans are the seat of spoken language, Broca's and Wernicke's areas. *H. habilis* was the earliest hominid to show both of these well developed.

We have therefore the revealing and provocative concurrence of several phenomena: the modern human sulcal pattern and the parts of the brain that govern spoken language became manifest at that stage when appreciable relative enlargement of the brain and marked encephalization first obtruded. As if that remarkable synchronicity were not enough, these major alterations in the structure of the brain became apparent at approximately the same time as deliberately fashioned stone tools first appeared in the archaeological record.

## THE DAWNING OF SPOKEN LANGUAGE

The human language capacity is an evolutionary achievement of 'extreme perfection and complication', to borrow Charles Darwin's phrase. What evidence do we have for the appearance, on brains and endocasts of living and fossil higher primates, of Broca's area and of the parieto-occipito-temporal complex (POT), including Wernicke's area?

First, in non-hominid primates, Geschwind, following the earlier work of Elliot Smith (1907) and of Critchley (1953), stated:

> The situation in man is not simply a slightly more complex version of the situation present in the higher primates but depends on the introduction of a new anatomical structure, the human inferior parietal lobule, which includes the angular and supramarginal gyri ...
>
> (Geschwind, 1965: pp. 273)

No trace of this inferior lobule is detectable in the macaque. It is present, though only in rudimentary form, in apes (Geschwind, 1965). Geschwind (1965) acknowledges that 'the exact degree of the uniqueness of the inferior parietal region in man remains to be determined' pp. 276.

There is scarcely any trace of impressions over the language-relevant cyto-architectonic areas in archaic apes before the appearance of the earliest hominids. Hence, it is not in the ancestral apes but in the earliest hominids of the genus *Australopithecus* that we might expect Broca's area and POT, or their immediate fore-runners, to have emerged.

When we seek more direct evidence, we are faced with the imperfection of the geological record. We have no good endocranial casts older than about 3 mya: in other words, the hominid brain is (thus far) mute for the first half of the time of humans on earth! *Australopithecus africanus* endocasts of the 3 to 2.5 mya period are small (scarcely larger, absolutely or relatively, than those of existing apes), have an essentially ape-like sulcal pattern (and the author agrees in this regard with Falk, 1983, 1989) and show slight development of a Broca's cap (Schepers, 1946) but no trace of inferior parietal lobule development.

Wilkins and Wakefield (1994) postulate that these areas were initially evolved to fulfil a function other than linguistic and they suggest that skilled manipulative activities constituted these non-linguistic functions. They suggest that these areas were subsequently redeployed — by what they call 'evolutionary reappropriation' — for linguistic purposes.

From their analytical standpoint, they have arrived at the same conclusion reached 21 years ago, namely that *Homo habilis* (and not modern *Homo sapiens*) was the first hominid to possess the neural basis for language (Tobias, 1975, 1981a,b, 1983).

Valuable evidence bearing on the concept world and intelligence of *H. habilis* is to be found in the technology and typology of the stone tools associated with *H. habilis* (Parker and Gibson, 1979; Gowlett, 1984; Wynn, 1981). Inferences have been drawn from archaeological living sites as to the social behavior and survival strategies of early tool-making hominids. These cultural and social analyses led to the inference that such complex patterns of behavior could be taught to the children of *H. habilis* only by a more efficient teaching mechanism than imitation, employed by apes (Tobias, 1981a). From the cultural and social aspects of the lifeways of *H. habilis* and from the testimony of the endocasts, the author has thus held for the last 14 years that *H. habilis* was the first linguistic primate (Tobias, 1980, 1981a, 1983, 1987, 1991, 1994). This hypothesis has since been supported by other researchers including Falk (1983) and Andrews and Stringer (1993).

Wilkins and Wakefield relate the evolution of language to alternative motor activities (1995). Jerison (1977, 1991) has offered an interesting speculative analysis which sees the initial evolution of language, not as a communication system, but as a supersensory system, 'From an evolutionary point of view', says Jerison, 'the initial evolution of language is more likely to have been as a supplement to other sensory systems for the construction of a real world. This would be consistent with the other evolutionary changes in mammalian neural adaptations and would not require the sudden appearance of an evolutionary novelty. The suggestion is that our ancestors evolved a more corticalized auditory sense that was coupled with the use of vocal capacities for which almost all living primates are notorious' (1977, pp. 55).

There are at least three competing views about the emergence and evolution of language, one which relates it to a prior, non-linguistic, motor function; one which connects it to a prior, non-communicative sensory function, the building of a world image; and one which sees language emerging as a vehicle of communication from the beginning. Testability is considered the hallmark of a good, rigorous hypothesis. It will test the ingenuity of palaeo-neurobiologists well into the 21st century to convert these three theories into testable hypotheses.

## SUMMATION

In the stages by which the presumably ape-like brain of the last common ancestor of man and chimpanzee was made over into the brain of humankind, it seems that relatively small changes accompanied the emergence of the earliest available and analyzable hominids, the australopithecines. These changes comprised a minimal increase in absolute and relative size and some limited reorganization of the overall anatomical structure of the brain. As for neurologically important changes in the brain, there is scarcely any evidence of surface alterations in the sulcal and gyral patterns and, what differences have been claimed, especially in regard to the prestriate visual areas of the cerebral hemispheres, are to say the least problematical.

However, major expansion of the brain and critical cortical reorganization were striking features of the change from *A. africanus* to *H. habilis*. These changes included notable augmentation of the cerebrum, strong lateral expansion of the parieto-occipital region, the appearance of a human-like sulcal pattern and the emergence for the first time of protuberances interpreted as the anterior and posterior speech cortices.

Thus, it is with the appearance of the *H. habilis* brain that a gigantic step was taken to a new level of organization in hominid brain evolution. The changes apparent in the advancement of the brain from an australopithecine ancestor to earliest *Homo* were perpetuated in ensuing stages of evolution of the brain in later species of *Homo*. Thus, as far as external morphology of the brain was concerned, that of *H. habilis* was a small and ancient adumbration of the brain of modern humans. In size, however, the brain of *H. habilis* had a long way to go, from a mean capacity of 640 cm$^3$ to one of 1350 cm$^3$ in modern humans. Some part of that doubling of brain-size is dependent on the increase in body size from the pygmoid habilines to the taller and heavier men and women of *H. erectus* and *H. sapiens*, on the principle that bigger people have bigger brains. But this factor does not account for all of the post-habiline increase. When body size is taken into account and relative brain size is estimated from *H. habilis* through *H. erectus* to *H. sapiens*, there is clear evidence that further encephalization occurred right up to the emergence of modern *H. sapiens*.

The advantages of bigger brains, such as more information storage capacity, more cognitive faculty, more behavioral plasticity, did not operate to produce once-off responses. The process continued for two million years: so we must suppose that the selective advantages continued to operate for that period of time — across taxonomic boundaries, geographic zones, cultural diversity, ecological radiations and behavioral variegation.

What advantages were conferred by increasing encephalization during hominid phylogeny? Many selection pressures have been proposed (see Gabow, 1977; Tobias, 1981a; Jerison, 1991). Essentially, the advantage of bigger brains is that they are capable of processing more information — and that seems to have conferred a decided evolutionary advantage within a lineage of encephalizing species, such as that of the hominids. Any new explanation that may be offered on hominid encephalization must cover not only the inferred rapidity of the change that has occurred, but also the persistence of encephalization.

It is hypothesized that the relative enlargement of the hominid brain was a mechanism by which enhanced adaptability might be furnished and on which natural selection could go to work, while adaptation, that is, concurrent adaptedness, was not sacrificed and could even have been improved.

This general proposition should apply to all mammalian lineages characterized by progressive encephalization. In the hominid lineage it has perhaps attained its pinnacle of evolution, as reflected by modern humanity's remarkable degree of encephalization. Along this lineage, the particular property 'secreted' by the expanding brain was the cognitive faculty, of such quality and degree as to generate culture and language.

Probably no more puissant forces have yet appeared on earth, in their capacity to potentiate adaptation and to widen dramatically the evolutionary flexibility of their possessors. The effects of hominid encephalization and increased flexibility of function should not be lost on any model of the human brain.

# REFERENCES AND SUGGESTED FURTHER READING

Editor: Models that propose a later development of language are presented in Noble and Davidson (1996). An overview of our human lineage and implications for cognition is provided by Mithen (1996), and the extraordinary story of the initial findings of these 'missing links' is described in Leakey and Hay (1992).

Andrews, P. and Stringer, C. (1993) The primates' progress. In *The Book of Life,* ed. S.J. Gould. London: Ebury Hutchinson, (pp. 219–251).

Benade, M. M. (1990) *Thoracic and lumbar vertebrae of African hominids ancient and recent: morphological and functional aspects with special reference to upright posture.* M.Sc. Dissertation. Johannesburg: University of the Witwatersrand.

Berger, L. R. (1994) *Functional morphology of the hominoid shoulder, past and present.* Ph.D. Thesis. Johannesburg: University of the Witwatersrand.

Critchley, M. (1953) *The Parietal Lobes.* London: Edward Arnold.

Dart, R. A. (1925) Australopithecus africanus: the man-ape of South Africa. *Nature,* **115,** 195–199.

Elliot Smith, G. (1907) A new topographical survey of the human cerebral cortex, being an account of the anatomically distinct cortical areas and their relationship to the cerebral sulci. *Journal Anat., **41,** 237–254.

Falk, D. (1983) Cerebral cortices of East African early hominids. *Science,* **222,** 1072–1074.

Falk, D. (1989) Ape-like endocast of 'ape-man' Taung. *American Journal of Physical Anthropology,* **80,** 335–339.

Galaburda, A. M. (1984) Anatomical asymmetries. In *Cerebral Dominance: The Biological Foundations,* N. Geschwind and A.M. Galaburda, eds. Cambridge, Mass. Harvard University Press, (pp. 11–25).

Gowlett, J. A. J. (1984) Mental abilities of early man: a look at some hard evidence. In *Hominid Evolution and Community Ecology*, R. Foley, ed. London: Academic Press.

Geschwind, N. (1965) Disconnexion syndromes in animals and man. Part I. *Brain,* **88,** 237–294.

Grüsser, O. J. and Weiss, L. R. (1985) Quantitative models on phylogenetic growth of the hominid brain. In *Hominid Evolution: Past, Present and Future,* P. V. Tobias, (Ed). New York: Alan R. Liss, (pp. 457–464).

Holloway, R. L. (1974) The casts of fossil hominid brains. *Scient. Am.* **231,** 106–115.

Holloway, R. L. (1983) Cerebral brain endocast pattern of Australopithecus afarensis hominid. *Nature,* **303,** 420–422.

Holloway, R. L. (1988) 'Robust' australopithecine brain endocasts: some preliminary observations. In *Evolutionary History of the 'Robust' Australopithecines,* F. E. Grine, ed. New York, Aldine de Gruyter, (pp. 95–105).

Jerison, H. (1963) Interpreting the evolution of the brain, *Hum. Biol.,* **35,** 263-291.

Jerison, H. (1977) Evolution of the brain. In *The Human Brain,* M.C. Wittrock et al., eds. Englewood Cliffs, N.J, Prentice Hall, (pp. 39–62).

Jerison, H. (1991) *Brain Size and the Evolution of Mind.* James Arthur Lecture. New York: American Museum of Natural History.

Jungers, W. L. (1985) *Size and Scaling in Primate Biology.* New York: Plenum.

Keith, A. (1915) *The Antiquity of Man.* London: Williams and Norgate.

Lande, R. (1985) Genetic and evolutionary aspects of allometry. In W.J. Jungers (Eds), *Size and scaling in primate biology*, (pp. 21–32), New York: Plenum Press.

Lashley, K.S. (1949) Persistent problems in the evolution of mind. *Quart. Rev. Biol.* **24,** 28–42.

Leakey, R. and Hay R. L. (1992) *Origins reconsidered.* London: Little Brown.

Leakey, L. S. B., Tobias, P. V. and Napier, J. R. (1964) A new species of the genus Homo from Olduvai Gorge. *Nature,* **202,** 7–9.

Mithen, S. (1996) *The prehistory of mind.* Thames: Hudson.

Noble, W and Davidson, I (1996) *Human Evolution, Language and Mind.* Cambridge University Press.

Parker, S. T. and Gibson, K. R. (1979) A developmental model for the evolution of language and intelligence in early hominids. *Behavioral and Brain Sciences,* **2,** 367–381.

Schepers, G. W. H (1946) The endocranial casts of the South African apemen. In The South African Fossil Apemen: The Australopithecinae, R. Broom and GWH Schepers. *Transvaal Museum Memoirs,* **2,** 153–272.

Tobias, P. V. (1967) *Olduvai Gorge, Vol. 2: The Cranium and Maxillary Dentition of Australopithecus (Zinjanthropus) boisei.* Cambridge: Cambridge University Press.

Tobias, P. V. (1968) The pattern of venous sinus grooves in the robust australopithecines and other fossil and modern hominids. In *K. Saller Festschrift Anthropologie and Humangenetik.* Stuttgart: Gustav Fischer Verlag, (pp. 1–10).

Tobias, P. V. (1970) Brain-size, grey matter and race — fact or fiction? *Am. J. Phys. Anthropol.,* **32,** 3–26.

Tobias, P. V. (1971) *The Brain in Hominid Evolution.* Columbia University Press.

Tobias, P. V. (1975) Brain evolution in the Hominoidea. In *Primate Functional Morphology and Evolution,* ed. R. Tuttle. The Hague: Mouton Publishers, (pp. 353–392).

Tobias, P. V. (1980) Homo habilis and Homo erectus: from the Oldowan men to the Acheulian practitioners. *Anthropologie,* **18,** 115–119.

Tobias, P. V. (1981a) *The Evolution of the Human Brain, Intellect and Spirit.* University of Adelaide, Adelaide, South Australia.

Tobias, P. V. (1981b) The emergence of man in Africa and beyond. *Philosophical Transactions of the Royal Society, London,* **292,** 43–56.

Tobias, P. V. (1983) Recent advances in the evolution of the hominids with especial reference to brain and speech. *Pontificiae Academiae Scientiarum Scripta Varia* **50,** 85–140.

Tobias, P. V. (1987) The brain of Homo habilis: a new level of organization in cerebral evolution. *Journal of Human Evolution,* **16,** 741–761.

Tobias, P. V. (1991) Olduvai Gorge, Vols. 4A and 4B. *The Skulls, Endocasts and Teeth of Homo habilis.* Cambridge: Cambridge University Press.

Tobias, P. V. (1992) The species Homo habilis: example of a premature discovery. *Ann Zool. Fennici,* **28,** 371–380.

Tobias, P. V. (1994) The craniocerebral interface in early hominids. In *Integrative Paths to the Past,* R.S. Corruccini and R.L. Ciochon, eds. Englewood Cliffs, N.J.: Prentice Hall, (pp. 185–203).

Wilkins, W. K. and Wakefield, J. (1995) Brain evolution and neurolinguistic preconditions. *Behavioral and Brain Sciences.*

Wynn, T. (1981) The intelligence of Oldowan hominids. *Journal of Human Evolution,* **10,** 529–541.

# Chapter 4

# THE MIND–BRAIN PROBLEM

Peter Slezak

School of Science and Technology Studies, Program in Cognitive Science
University of New South Wales

The problem of explaining the mind persists essentially unchanged today since the time of Plato and Aristotle. For the ancients, it was not a question of the relation of mind to brain, but the question was fundamentally the same. For Plato, the mind was conceived as *distinct from the body* and was posited in order to explain knowledge which transcends that available to the senses. For his successor, Aristotle, the mind was conceived as intimately *related to the body,* as form is related to substance. On this conception, the mind was an abstract property or condition of the body itself, 'enmattered formulable essence', being always embodied in some material substance. In these two accounts of the mind–body relation we see views which represent the range of conceptions ever since.

## PERCEIVING A STAR — FEELING A PAIN

In modern times, the problem of understanding the nature of mind was eloquently articulated by the great neurophysiologist Sir Charles Sherrington who contrasted the central phenomenon with those in the domain of physical science:

> The energy-concept … embraces and unifies much. …Immense as it is, and self-satisfying as it is, and self-contained as it is, it yet seems but an introduction to something else.
>
> For instance a star which we perceive. The energy-scheme deals with it, describes the passing of radiation thence into the eye, the little light-image of it formed at the bottom of the eye, the ensuing photo-chemical action in the retina, the trains of action-potentials travelling along the nerve to the brain, the further electrical disturbance in the brain, the action-potentials streaming thence to the muscles of eye-balls and of the pupil, the contraction of them sharpening the light-image and placing the best seeing part of the retina under

it. The best 'seeing'? That is where the energy-scheme forsakes us. It tells us nothing of any 'seeing'. Everything but that. Of the physical happenings, yes. … But, as to our seeing the star it says nothing. … The energy-scheme deals with the star as one of the objects observable by us; as to the perceiving of it by the mind the scheme puts its finger to its lip and is silent. It may be said to bring us to the threshold of the act of perceiving, and there to bid us 'goodbye' (1942, pp. 304–5).

For Sherrington, the fundamental aspect of mental life is conscious experience which physical science cannot explain. It is striking, but no accident, that over fifty years later, discussions of the problem echo Sherrington. Thus, leading philosopher John Searle invites the reader to perform a small experiment in order to illustrate the essential nature of consciousness, suggesting that the reader pinch the skin of an arm. Then, like Sherrington, Searle enumerates various facts about the sequence of neuronal events from receptors to cortex, but adds:

A few hundred milliseconds after you pinched your skin, a second sort of thing happened, one that you know about without professional assistance. You felt a pain. If you were asked what is the essential thing about the sensation of pain, I think you would say that the second feature, the feeling, is the pain itself. The input signals cause the pain … But the essential thing about the pain is that it is a specific internal qualitative feeling. The problem of consciousness in both philosophy and the natural sciences is to explain these subjective feelings (1997, pp. 99).

## FAMILIAR BUT ELUSIVE: THE MYSTERY OF CONSCIOUSNESS

The ancient problem of relating mind to body, like the modern one, has centered upon the phenomenon of consciousness. The question of consciousness is among the greatest mysteries of all. Indeed, Chalmers (1996) adds that it may be the principal obstacle in the quest for a scientific understanding of the universe. Chalmers characterizes the 'easy problems' of mind as those concerning the various psychological and neurological phenomena — not in the sense that they are intellectually trivial, but only in the sense that they are tractable to scientific explanation. By contrast, however, he characterizes the 'hard problem' as that of subjective experience which generates a notorious 'explanatory gap'. This is a gap between objective accounts of physical, causal or functional phenomena on the one hand, and subjective accounts of first-person, phenomenal or qualitative experience on the other. As Block (1997) notes, 'At this stage in the relevant sciences we have no idea how the neural substrate of my pain can explain why my pain feels like this rather than some other way or no way at all.' pp. 175. Whether or not this explanatory gap can be bridged constitutes the core of the mind–brain problem.

The fundamental perplexity has been aptly captured by McGinn, who asks: 'How is it possible for conscious states to depend upon the brain? How can Technicolor phenomenology arise from soggy grey matter?' (McGinn, 1991: pp. 1). The spectacular advances in neuroscience are as remarkable for their failure to bridge this gap as for the insights they have undeniably provided. The intractability of this puzzle has led

McGinn to the conclusion that it is insoluble in principle. Like Sherrington, other philosophers, too, such as Chalmers (1996) and Searle (1997) argue that even the most complete, utopian neuroscience would leave essential features of mind and consciousness unexplained.

## CONSCIOUSNESS EXPLAINED?

On the other hand, there is an alternative view also widely held nowadays among philosophers and cognitive scientists. While there undoubtedly remain vastly many facts about the mind and brain which are not yet understood, there is nevertheless some justification for the view that the fundamental mind–brain problem has been solved after all. It is in this vein that Dennett (1991) boldly proclaims that we now have *Consciousness Explained* — to use the title of his influential book. Such claims receive justification partly from scientific advances in the interdisciplinary field of 'cognitive science', including neuroscience and artificial intelligence (AI). On the one hand, significant mental abilities ranging from problem-solving to vision have been produced in computers, leading to the claim of 'Strong AI' that mind is just symbolic computation. On the other hand, much of neuroscience itself has been cast in computational, information-processing terms, reinforcing the conception of mind as computational. More broadly, however, a conviction that the mind–brain problem has been solved in principle is based on a scientific world-view which places human beings within the natural order and, therefore, completely explainable by the laws of physics and chemistry. Whatever might be the peculiarities of the mind, these must be products of the complex causal processes of the brain. This general view known as 'materialism' is a commitment to the adequacy of natural science to explain the mind in the same terms as it explains the rest of the non-mental world. Searle (1997) charges Dennett with simply denying the very existence of the phenomena of conscious mental states (the 'first person' introspective qualities of experience) to be explained, thereby evading rather than solving the problem.

The apparent recalcitrance of subjective states or 'qualia' of conscious experience has been taken as evidence for the inadequacy of any scientific account including the currently favored computational or functionalist versions. For example, two systems might be materially or functionally equivalent in every respect and yet still differ in the experienced qualia: one person's subjective experience of colors might be systematically reversed from that of another, though they might be physically or functionally identical and there would be no objective way to tell the difference. Thus, what one person sees as blue might, in fact, be what another person sees as green, but their use of color, words and all other behavioral tests could not discriminate between them. Likewise, it is argued that a robot might be functionally or physically indistinguishable from human beings but yet fail to have any 'inner', subjective, phenomenal experience at all. The alleged possibility of such cases has been taken to demonstrate the inadequacy of purely physical or functional models as a complete explanation of the mind.

# DUALISM

## The Transcendental Temptation

Materialism is contrasted with Dualism which has been the traditional view of the mind as a non-physical, immaterial or spiritual substance. Notoriously associated with the 17th Century philosopher René Descartes, dualism takes the mind entirely out of the realm of scientific explanations altogether. Remarkably, there has been a consensus among some leading neuroscientists on a dualist solution to the classical mind–brain problem, including Sherrington (1942) in *Man on his Nature*, Penfield (1975) in *Mystery of the Mind*, and Eccles (1970). It is a curious anomaly that the view endorsed by these eminent neuroscientists is essentially Descartes' dualism of body and soul.

Dualism is the view characterized in Bertrand Russell's quip: 'What is mind? No matter. What is matter? Never mind!'. In other words, dualism is the view that there are two fundamental kinds of entities or substances in the world, namely, physical objects and immaterial minds or perhaps souls. It is roughly the view most people get in Sunday school. Although this doctrine is certainly the most widely held view among laymen, it is nowadays dismissed by most philosophers as not merely implausible, but without scientific merit and, therefore, no longer deserving to be taken seriously among the contending theories of mind. To be sure, Descartes also held this view, but that was three hundred years ago and he, even then offered subtle and profound arguments based on the scientific considerations of his time.

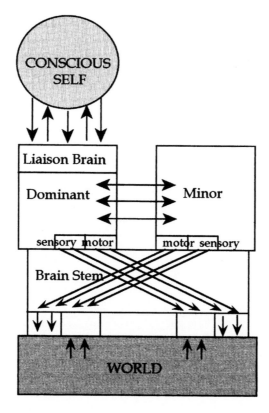

**Figure 4.1.** Dualist model of mind–brain interaction (after Eccles, 1970).

At first sight, nobel laureates in neurophysiology who advocate dualism appear to be an embarrassment to philosophers who propound the orthodox materialist line. After all, who can claim more authority on the the relation of mind and brain than such distinguished neuroscientists? In fact, however, the arguments which purport to be based on the evidence of neuroscience are surprisingly inadequate by the standards of scientific reasoning. Of course, from an explanatory point of view, the resort to a non-physical realm is a curious tactic, being perhaps more comforting than cogent. In the absence of any specific details concerning the manner in which an immaterial substance might perform the special tasks, it is rather vacuous as an explanation. Typically no details are offered concerning the way in which the phenomena of consciousness could be produced by non-physical processes and therefore there is no explanatory advantage in transferring our ignorance to such another realm — quite apart from its other serious drawbacks.

Worse than being merely gratuitous, the dualist position is intellectually costly because it comes into conflict with the most fundamental principles of science, including deeply entrenched principles of physics. If the 'conscious self' of Eccles can interact with the brain via the the alleged 'liaison' regions (see figure 4.1), then there must be some physical effects arising from non-physical causes — that is, some energy transactions arising from nothing. This 'paranormal' or miraculous character of the claim has always been the source of its most serious implausibility[1].

Eccles has a propensity to dress up specious arguments with the latest developments in neurophysiology, but this is misleading because, in fact, the science provides no support at all for his claims. Thus, for example, Eccles gives a brief account of the columnar, modular structure of the cerebral cortex, referring to Mountcastle's pioneering work, and then he asserts

> The simplest hypothesis of brain–mind interaction is that the self-conscious mind can scan the activity of each module of the liaison brain or at least those modules tuned in to its present interest (1980, pp. 46).

However, it seems clear that it would not matter what the neurophysiological facts were, since they play no substantive role in Eccles' account and provide no explanatory support of any kind for his dualist interactionism. Clearly Eccles could tell exactly the same story about the 'liaison brain' whatever the neurological facts were.

---

1   We see Eccles (1980, pp. 19, 20) embrace the conflict between dualism and physical science, asserting triumphantly that 'nowhere in the laws of physics or in the laws of the derivative sciences, chemistry and biology, is there any reference to consciousness or mind' (1980, pp. 20). This is, of course, quite correct, but Eccles is mistaken in seeing this as support for dualism. The situation he notes is precisely typical when phenomena are supervenient upon, or theoretically reducible to, others and, in this sense, 'emergent'. Thus, the property of wetness does not appear in the theories of hydrogen and oxygen despite being an emergent property of their combination as water. Likewise, fundamental physics does not mention economic facts such as inflation, but we would hesitate in concluding that they must transcend the known world of material phenomena. In the classical text-book examples of theoretical reduction, such as that of thermodynamics to statistical mechanics, we have the independent principles of one discipline derivable from the laws of another. However, the concepts of the reduced domain do not themselves appear in the reducing one. Here, too, one would hardly suggest that temperature and pressure must be somehow outside the bounds of physical science, and hence inexplicable in these terms, on the grounds that these concepts do not appear in more basic laws of physics.

## René Descartes: The Ghost in the Machine

Descartes is a pivotal figure in the history of modern philosophy and modern science, and his work serves as a valuable point of reference for understanding the contemporary scene. Above all, the recent move in philosophy of mind towards a deep engagement with neuroscience makes it easier to appreciate Descartes' contribution. Desmond Clarke (1982) has said ironically that Descartes must be understood as 'practicing scientist who, somewhat unfortunately, wrote a few short and relatively unimportant philosophical essays' pp. 2. However, the obsolescence of Descartes' scientific work meant that it had been largely ignored among 20th Century philosophers. Nevertheless, Descartes' neglected researches showed remarkable insights and anticipations of current views. For example, he understood that the path from the retina to the brain involved the encoding of information which he illustrated in his *Traité de L'homme* by the analogy of a blind man's cane. Nothing physical is conveyed along the cane but only signals encoding information about the objects touched.

Descartes is not only the source of a dualism parodied 'with deliberate abusiveness' by Gilbert Ryle (1949) as the doctrine of the 'ghost in the machine', but also an important contributor to the mechanical world-view. Indeed, Descartes' dualism is better understood as a rather insightful consequence of his thoroughgoing mechanistic explanation of the human body. Despite seeing the human body as an exquisite machine, Descartes perceived the finite limitations of clockwork mechanisms (the leading-edge technology of his time) and he was led to postulate some other basis for certain human abilities which could transcend the limitations of clockwork mechanizing.

Descartes recognized that the fixed repertoire of any mechanical, clockwork contrivance could not explain the complex features (language, acting through knowledge) of human behavior. No matter how many gears and levers in a clockwork machine, there is, indeed, only a finite number of behaviors it is capable of executing, whereas humans appear to act spontaneously in ways which are unanticipated and appropriate to novel circumstances. It is important to appreciate that Descartes was right about this, for it is only with the work of Alan Turing in the Twentieth Century that we have an entirely different conception of machines, namely, computers which transform, not work, but information. While no less 'mechanical' or physical, such machines are capable of novelty or creativity precisely because they are designed to represent and act through knowledge. Thus, in marked contrast with more recent versions of dualism, Descartes' reasons for postulating a non-physical mind were arguments based on deep insights and the limited scientific knowledge available to him.

Quite independent of this argument, Descartes had a different and better-known argument for the existence of an immaterial and immortal soul. This is the argument concluding with the famous dictum *cogito ergo sum* — I think, therefore, I am. The vast literature on this argument over three hundred years has failed to clarify its character, its logic or its supposed force, but the intricacies of this question cannot be entered upon here. It must suffice to note only that the argument depends on the apparent certainty of self-knowledge by contrast with the doubtfulness of everything else. For Descartes, this suggested the independence of the mind from the rest of the material world. He could doubt everything except his own thoughts. This argument is also deeply compelling, though it cannot be regarded as a convincing proof of dualism.

Nevertheless, it continues to be a subject of analysis and controversy in the philosophical literature (see Slezak, 1983).

## Behaviorism

For its founder J.B. Watson (1913), Behaviorism was explicitly conceived as an attempt to place psychology on a rigorous, objective, scientific basis in order to avoid Cartesian dualism, which he identified with religious doctrines of the soul. The hoped-for scientific objectivity was to be achieved by confining psychology to strictly observable causes and effects — the stimuli and responses of the organism. Behavior was assumed to be caused by stimuli and psychology was conceived as the science of stimulus–response correlations. In particular, all talk of mind, consciousness, qualia and other inner mental processes was to be eschewed as unscientific.

For Skinner (1938) such postulation of inner processes was seen as a vacuous pseudo-explanation because he thought it led to an infinite regress. With some justice, Skinner saw the postulation of inner processes as merely explaining intelligence with intelligence and, therefore, begging the question. This is the notorious problem of positing an 'homunculus' or 'little man' in the head which is empty from an explanatory point of view. For example, if visual perception is explained in terms of an image transmitted from the eyes to the brain, it would seem to require another internal pair of eyes to see this image and so on *ad infinitum*. However, as Dennett (1978) has shown, in his concern with this problem, Skinner was partly right and partly wrong. The dilemma seems to be, as Dennett puts it, that psychology *with* homunculi is impossible because of the notorious regress, but psychology *without* homunculi is also impossible. Psychology without homunculi — that is, a behaviorist psychology — was sterile in practice and shown to be bankrupt in principle (Chomsky 1959).

The fear of postulating the mind was derived from the misguided fear of unobservable, inner entities which, in turn, arose from a failure to understand the essential feature of any fruitful science. In physics, for example, Behaviorists' own model of successful science, there is no effort to avoid the postulation of unobservable objects. As Chomsky has wryly observed, if physics were to be practiced according to behaviorist strictures, it would become the science of meter readings. Chomsky's meticulous analysis revealed that the pretensions of behaviorism to being objective and non-mentalistic were a sham. The stimulus-response correlations turned out to make a covert appeal to the very inner processes supposedly being avoided. That is, the allegedly objective identification of external causes and effects merely disguised assumptions about internal processes. In this, as in other respects, as Chomsky noted, behaviorism adopted only the superficial trappings of science without its deeper explanatory content.

## MATERIALISM
### From Identity Theory to Functionalism

No longer concerned to refute dualism, numerous philosophers in this century (Place, Smart, Armstrong) attempted to formulate the precise sense in which the mind was to be identified with the brain. The details of neuroscience were not thought to be relevant to this concern, since the problem was only to formulate the mind–brain identity in a way which would be satisfactory as a general thesis. Without entering into the subtle details, we may note that materialism was conceived in the form of an Identity between brains and minds, taken to have the status of a contingent, scientific fact just like the identity of water with $H_2O$, or lightning with electrical discharge. However, despite the initial plausibility of such an identification, it soon became evident that there were certain difficulties in formulating this *Identity Theory*.

In one form, known as type-type identity, the theory makes the sweeping claim that all the mental events which could possibly exist must be identical with neurological events. However, it seemed clear that this version of materialism was too restrictive, since it ruled out in principle the very possibility of mental events being discovered in anything other than brains. In particular it was important to leave open the possibility that minds might emerge not only in brains but also in radically different things such as computers. Such possibilities cannot be ruled out in principle and they suggest that we must not identify minds strictly with any particular material substrate in which they might happen to be embodied.

The consequence of these considerations was that the mind must be identified with something more abstract than its underlying substrate — just as a piece of music is not identified with the CD, tape or vinyl on which it happens to be recorded.

Pattern and organization are not some ethereal, spiritual substance but only the abstract information encoded in some form. In particular, the mind was therefore not to be identified directly with the brain as such, but rather with the information encoded in the brain, thereby permitting the multiple realizability of the relevant patterns. The relation of mind to brain on this view was analogous to the relation of software to hardware in a computer. In particular, and ironically, this Functionalist account of the mind was now consistent with Cartesian dualism since the possibility of minds being realized in soul-stuff was no longer ruled out in principle. Instead, the implausibility of dualism now arose only from the unlikelihood of there being any such spiritual substance as a general scientific matter.

In sum, then, the problem of identifying the mind strictly with the brain was that the mental constitution of a system seems to depend, not on its specific constitution or composition, but rather upon its organization. The point is that there seems to be a level of abstraction at which the generalizations of psychology or mental attributions are most naturally formulated. In this sense, the functionalist account is not only a solution to the metaphysical mind–brain problem but also a methodological precept for the practice of cognitive science.

## REDUCTIONISM, LEVELS OF ANALYSIS AND MODELING THE BRAIN
### Reductionism

For some philosophers of mind, the impossibility of reductionism has meant the irrelevance of neuroscience to an autonomous science of the mind, since the brain is merely the 'hardware' substrate or physical realization of the abstract informational states, and it is only the latter, the 'software', which is relevant to understanding the mind.

Thus, Fodor (1981) suggests that the reason psychology is unlikely to reduce to neuroscience is analogous to the reason that economics is unlikely to reduce to, say, physics: the category of monetary exchange, for example, is unlikely to be co-extensive with any physical category. Likewise, psychological categories might be realized by a wide variety of mechanisms both among different organisms and even within the same organism and, therefore, unlikely to bear the appropriate deductive logical relation to neurological concepts.

It is crucial to notice that this kind of skepticism about the prospect of a reduction of mind to neuroscience is not an ontological claim. The impossibility of a theoretical reduction of psychology to neuroscience is consistent with the brain being the physical basis for the mind. That is, even if particular mental events are identical with neurological events, Fodor explains, 'it does not follow that the kind of predicates of psychology are coextensive with the kind of predicates of any other discipline' (1981, pp.135). Similarly, even if particular monetary exchanges involve transferring round metal tokens, it does not follow that the former concept corresponds precisely with the latter behavior in general. The issue of reductionism is the issue of relations between laws and concepts of different disciplines and whether they bear appropriate logical relations to one another.

## Levels of Analysis

Among the most influential works in cognitive science has been David Marr's (1982) *Vision,* whose computational account is of particular interest here because it emerged directly from the problems confronted in modeling the brain. Marr cites pioneering work in neuroscience such as the single cell studies of Hubel and Wiesel (1965) as holding out the promise of an understanding of the brain as a whole. He says 'There seemed no reason why the reductionist approach could not be taken all the way' (1982, pp. 14). However, despite the exciting advances at this level of inquiry, Marr suggested that something was wrong as the initial discoveries of the 1950s and 1960s were not followed by equally dramatic progress. Researchers were turning away from the original kinds of inquiry and Marr suggests that the problem was that 'None of the new studies succeeded in elucidating the function of the visual cortex' (1982, pp. 15).

This concern with the function of the brain identifies a crucial level of analysis in Marr's tripartite scheme. This highest or most abstract level of analysis is concerned with what a system does and why, and is referred to by Marr as the computational theory of the system, which specifies a mapping of one kind of information to another. In the case of vision, the task of a theory at this level is to derive properties of the world from images of it. Marr suggested that understanding the computational nature of the problem being solved by the system is an important pre-requisite for discovering what is going on at other levels of analysis such as that of the 'hardware' implementation in the brain.

According to functionalism, then, psychological states are to be characterized in terms of the causal roles they play in the overall system of states that mediate between sensory input and behavioral output. By analogy, a camshaft in an engine is identified in terms of its causal role — the 'whatchamacallit' that rotates the 'whatsits' that makes the 'thingamajig' go up and down. The point is that *anything* which satisfies this functional description in terms of its role in the system counts as a camshaft, regardless of its material composition. A specification of mental states in this way, too, is by reference to their relational properties at an abstract level of analysis and not by reference to their material instantiation. Thus, more prosaically, for example, the notion of 'mousetrap' is a functional description, since it can be implemented or realized in any number of disparate material devices: spring traps, cages or even cats.

Once functionalism is spelled out along these lines, as Dennett (1978) explained, the interest of artificial intelligence research to modeling the mind–brain becomes clearer. Computer programs can be regarded as an alternative formalism for stating psychological models — with the particular advantage that the merits of the model are more easily determined by seeing whether the program works. Thus, a theory of chess playing or language when formulated as a computer program will actually play chess or understand language. In this way computer programs can be construed literally as theories, and AI as a species of top-down cognitive psychology (see comment on ANNs in The Big Picture and computer models chapters).

Dennett (1988) has noted that perhaps the most widely held prejudice about explaining the mind has been the view that psychology could turn out to be like physics, in the sense that it would reveal universal generalizations in terms of deep, comprehensive principles or laws of nature. This is the functionalist conception we have seen which entails a partition of explanatory levels, with the computational level as the appropriate one for expressing the physics-like laws of thought. However, an alternative possibility is that psychology is more like engineering than physics, because the mind is more like a complex gadget or contraption produced by tinkering. In this case, Dennett points out that the principles we are seeking must be sensitive to pragmatic questions of design in a real world rather than those of an ideal designer or divine artificer. Evolution has been thought to operate in this manner as tinkerer rather than ideal designer (Gould 1980), and the virtues of connectionist models, for example, especially when most biologically faithful models may derive from their being more akin to such pragmatically engineered devices. In particular, there is no meaningful distinction between software and hardware in the case of connectionist models and this collapsing of explanatory levels is a feature which neural nets share with dynamical systems.

Indeed, there has been a move towards dynamic models of the mind which are not conceived in terms of the laws and levels of the computational account. Such models have been stimulated and reinforced by various criticisms of the computational view, particularly its commitment to formal symbolic representations as an internal 'language of thought'. Notorious among these critiques has been John Searle's (1980) attack on 'Strong AI' and its underlying functionalist conception of the mind as the program of the brain. Before turning to Searle, however, it is worth noting another source of dissent from the computational view of the mind.

## Mind as Algorithm of the Brain: Gödel's Theorem

In a suitably subtle and sophisticated form, the idea that the mind is the 'program' of the brain has been the foundational doctrine of the cognitive revolution and the inspiration for artificial intelligence, as well as information-processing psychology and neuroscience. However, grounds for challenging this computational view of the mind have been seen in the foundations of mathematics and the theory of computability in the form of Gödel's Incompleteness Theorem. Briefly, this source of doubt about the computational model of mind arises from the fact that a computer is essentially the embodiment of a formal axiomatic mathematical system. However, such formal systems were dramatically revealed by Kurt Gödel in 1931 to have certain inherent and insuperable limitations. Although mathematical truths were previously thought to be those statements which had formal proofs in some axiomatic system, Gödel revealed that there are, after all, some statements which are unprovable in principle but are, nonetheless known to be true. In this sense, a wide class of formal systems is said to be 'incomplete'.

Since Gödel showed that the sentences which are unprovable in some formal system can be known by us to be true, this suggests that there are some things which human minds can do but which computers cannot. This would be an inherent limitation on any machine which is a realization of a formal system and, thereby, evidence that the mind is not explainable as a computer of any sort. This argument was considered briefly by Alan Turing (1950) and developed in detail by Lucas (1961). Lucas concluded that Gödel's Theorem demonstrates that no algorithmic machine can be an adequate model of the mind, that minds are essentially different from any computer. Lucas's argument is open to serious objections, although it has been revived by the physicist Roger Penrose (1989) who takes human creative insight to be the essential feature of mentality which cannot be explained in terms of formal, algorithmic computation. It is unlikely that the phenomenon of insight should operate so far outside the ordinary physical, causal mechanisms of the brain as to require radically alternative and unknown explanatory principles (see Dennett 1989). While not a dualist approach as such, Penrose's resort to such principles outside the boundaries of currently known science seems gratuitous and reminiscent of Eccles' strategy we have noted.

## Searle's Chinese Room

Any survey of recent discussions of mind and brain must include some mention of the extraordinarily persistent debates surrounding John Searle's conundrum of the 'Chinese Room'. First articulated in 1980, Searle's argument has particular interest here in view of its assertion of the intrinsic dependence of the mind on the biological properties of brains. In particular, Searle's argument is an attack on functionalism and 'Strong AI' which envisions genuine consciousness in machines. Searle believes that the phenomena of 'intentionality' are the special products of brains which cannot be reproduced by computations on symbols. Intentionality is the philosophical term for the property of mental states which makes them have representational content or, in other words, to be be about something. The particular interest of computational views of the mind seemed to be the prospect of capturing this representational nature of mental states, since computers are important precisely through their ability to encode, store and manipulate information. However, the abstract, formal, syntactic nature of such symbolic representation, seems not to capture the full meaningfulness of mental representations. Instead, Searle suggests that intentionality must be understood as a biological, causal property of brains, intrinsically dependent on the biological substrate.

Searle's challenge was to the foundational assumption of AI and cognitive science: namely, the Physical Symbol System Hypothesis of Newell and Simon (1976). This asserted the equivalence of mind with the manipulation of abstract, formal symbols —

taken to be essentially meaningless tokens manipulated according to certain rules. The intrinsic meaninglessness or purely formal, syntactic character of the symbols in a computational model, makes it difficult to explain how the intentional, semantic content or 'aboutness' of mental representations could be explained. For this reason, Searle sought to refute the identification of mind with symbolic computation — by posing a thought-experiment, the famous Chinese Room puzzle.

Essential to Searle's Chinese Room scenario is an analogy between the meaningless symbols of a computational model and the meaningless squiggles of Chinese characters to an English speaker. Searle imagines that he is inside a room furnished with a rule book which correlates Chinese characters with other Chinese characters. Although Searle does not understanding any of these characters, he uses the rule book to communicate with a Chinese speaker outside the room. Thus, from his own point of view, he receives meaningless squiggles on pieces of paper through a mail slot and sends other meaningless squiggles back. However, from the point of view of the Chinese speaker outside the room, he is conducting an ordinary, meaningful dialogue which is indistinguishable from the responses of a fully intelligent person. Searle insists that there is nothing 'inside' which corresponds to the intelligibility or genuine meaningfulness which normally accompanies conscious awareness. Above all, the story can be seen as posing a skeptical challenge to computational accounts of the mind, for it appears that the human brain and its representational states have intrinsic meaningfulness and not merely arbitrary interpretations imposed upon formal symbols.

Searle's argument rests, in part, on a distinction between mere simulation and the real thing. Thus, a computer-simulated hurricane does not cause papers on the desk to get wet and blow away. Analogously, Searle argues that a simulation of human thought, no matter how complete or realistic, is still not genuine thought with the rich internal meaningfulness of the representational states. In short, Searle allows the most complete success of AI in modeling the functional, computational processes of the mind, but insists that these can never actually constitute a mind like our own. However, *prima facie*, at least, there is an important difference between the usual cases of complex simulations and AI programs. In the former case, the internal symbols are taken to represent certain external, real-world phenomena such as the air-pressure or temperature. However, computer models of psychological tasks are different in significant ways. In these cases, the phenomena being modeled are themselves information or symbolic process. Thus, a computer simulation of problem solving actually solves problems. For the same reason it is small consolation to protest that my personal computer chess program doesn't really win when it beats me, but only simulated beating me! In such cases, unlike the weather, the task being modeled is itself essentially an information processing one, and so the distinction between simulation and the real thing collapses. The computer did not merely simulate beating me, but really beat me.

If we take the Chinese Room thought experiment as an appropriate method for comparing brains with computers, then the crucial question should be what a brain would look like from the same perspective Searle uses to examine an AI program — that is, metaphorically from within the Chinese Room. Searle does not ask this question and instead makes a misleading contrast between the Chinese symbols and

English ones in the room. However, if the Chinese characters are supposed to be metaphors for internal computational symbols, then the appropriate comparison and parallel perspective would be the view from somewhere inside the brain. The relevant comparison is surely between what goes on inside a computer and what goes on inside a brain. But, of course, from this vantage point, the processes at any level in the brain would appear just as meaningless as the 'Chinese' symbols inside a computer.

Whether we chose to examine action potentials, neurotransmitter release, patterns of activation across systems of neurons or any other mechanism within a brain, we would perceive nothing meaningful in the brain either. Thus, when applied consistently, Searle's method does not distinguish the inner workings of a computer from those of a brain. Searle's mistake is his assumption that meaningfulness to an intelligent observer-homunculus is a relevant or appropriate criterion to judge whether a system has genuine intentionality. Searle's requirement that the internal symbolic representations should be intelligible when read by an observer entails that, in the case of brains having genuine intentionality, the 'language of thought' must be English. This implication produces a *reductio ad absurdum* of Searle's argument and reveals that his invocation of unspecified 'causal powers' is gratuitous as explanation of genuine intentionality.

## Dynamic Models

The Big Picture section of this book and all the chapters on brain models serve to present the current status and future directions of modeling brain and mind. The view that mind cannot be explained by formal symbol manipulation is shared by Walter Freeman (Freeman and Skarda 1991 and Freeman's chapter 10b, Brain Chaos and Intentionality) who propose a dynamical view of brain function which is not only taken to challenge the symbolic computational view, but is also offered as a radical departure from conventional models of brain function in terms of individual neurons and their actions. This research has 'philosophical' interest through appearing to offer a solution to the vexed puzzle of intentionality, providing a specific neurological basis for the meaning or content of thoughts. Though claiming to repudiate internal representations altogether, these claims are best understood as rejecting a certain particular conception of representations: namely, the orthodox, symbolic computational account. Indeed, their own models of self-organized, spatially coherent, stimulus-specific patterns of activity may be seen as constituting the intentional content of internal representations which has been sought by philosophers during decades of intense controversy.

Regarding the computational view of mind, Freeman's complaint is that 'Evidence from our laboratory indicates that brain functioning does not resemble the rule-driven symbol-manipulating processes characteristic of digital computers' (Freeman and Skarda, 1991: pp. 116). Although claiming to agree with Searle's skepticism about symbol manipulation and computation, in a different context Skarda and Freeman diagnose exactly the source of Searle's problem: they observe that 'physiologists are not so naive as to conclude, because they don't find symbols floating around in the neural tissue, that the brain is not a symbol-manipulation system' (Skarda and Freeman, 1987: pp. 184-5). As we saw, it is just this spurious source of skepticism which is implicit in Searle's Chinese Room scenario.

## CONCLUSION

At first sight, a chapter on the classical Mind–Brain problem may seem anomalous and premature in a book devoted to Integrative Neuroscience and approaches to modeling the brain. However, we have seen that, despite the sophistication of contemporary accounts, classical perplexities have persisted essentially unchanged at the heart of scientific theorizing itself.

Even if not bridging the 'explanatory gap' between mind and brain, undeniably, neuroscience has been increasingly seen as offering important new insights into the perennial philosophical conundrum. Conversely, philosophy, for its part, has become intensely engaged with empirical and theoretical developments concerning the brain. In this way, philosophy of mind and neuroscience are more intimately connected today than at any other time since the seventeenth century. Thus, the Mind–Brain problem in its modern version is no longer confined to the puzzles of consciousness, but emerges from substantive developments in the neurosciences, cognitive psychology, linguistics and AI. Despite the persistent concern with the ancient, seemingly intractable, puzzles of consciousness, much philosophical inquiry has simply abandoned the old questions in favor of an entirely different kind of enterprise. In the words of Patricia Churchland and Terrence Sejnowski (1990, pp. 228) 'the mind–body problem in its contemporary guise is this: Can we get a unified science of the mind–brain? Will psychological theory reduce to neuroscience?' Thus, it is significant, if not exactly the norm, that a philosopher such as Churchland has been co-author of a work entitled *The Computational Brain* (Churchland and Sejnowski, 1992) which is a technical work having no resemblance to the philosophy of a generation ago.

Other chapters in this volume make important contributions to this vision of a unified science, while the present chapter has sought to give some background in order to show how traditional questions have become recast in the modern context. There is a pleasing irony in the fact that philosophical discussions of the mind in the modern period begin with Descartes, since he is not only notorious for the dualism from which much of modern philosophy has departed, but he was also deeply engaged in studies of the brain to which it has returned. For Descartes at least, the recent 'innovations' of 'Neurophilosophy' would have been *déjà vu*.

## REFERENCES AND SUGGESTED FURTHER READING

Block, N. (1997) Begging the Question Against Phenomenal Consciousness. In N. Block, O. Flanagan & G. Gÿzeldere, eds. *The Nature of Consciousness: Philosophical Debates*, Cambridge, Mass., Bradford MIT Press, 175–179.

Chalmers, D. (1996) *The Conscious Mind: In Search of a Fundamental Theory*, Oxford, Oxford University Press.

Chomsky, N. (1959) Verbal Behavior, by B.F. Skinner. *Language,* **35**, 26–58.

Churchland, P. S. (1997) Can Neurobiology Teach Us Anything about Consciousness? In N. Block, O. Flanagan & G. Gÿzeldere, eds. *The Nature of Consciousness: Philosophical Debates*, Cambridge, Mass., Bradford MIT Press, 127–140.

Churchland, P. S. and Sejnowski, T. (1992) *The Computational Brain*. Cambridge MA. MIT Press.

Churchland, P. S. and Sejnowski, T. (1990) Neural Representation and Neural Computation, In W. Lycan ed. *Mind and Cognition*, Oxford, Basil Blackwell 224–252.

Clarke, D. M. (1982) *Descartes' Philosophy of Science*, Manchester, Manchester University Press.

Dennett, D. C. (1991) *Consciousness Explained*. Boston: Little, Brown and Co.

Eccles, J. C. (1970) *Facing Reality*, New York, Springer-Verlag.

Eccles, J. C. (1980) *The Human Psyche*, New York, Springer-Verlag.

Fodor, J. (1981) Special Sciences In his *RePresentations: Philosophical Essays on the Foundations of Cognitive Science*, Cambridge, Mass.: Bradford/MIT Press, 127–145.

Freeman W. J. and C. A. Skarda (1991) *Mind/Brain Science: Neuroscience on Philosophy of Mind*, in E. Lepore and R. Van Gulick eds. John Searle and his Critics, Oxford, Blackwell, 115–127.

Gould, S. J. (1980) *The Panda's Thumb,* Harmondsworth: Penguin.

Lucas, J. R. (1961) Minds, Machines and Godel, *Philosophy*, **36**, 113–117.

Marr, D. (1982) *Vision*, New York, W.H. Freeman.

McGinn, C. (1991) *The Problem of Consciousness,* Oxford, Blackwell.

Newell, A and H. A. Simon (1976) *Computer Science as Empirical Inquiry*: Symbols and Search, Communications of the Association for Computing Machinery, pp. **19**, 113–126.

Penfield, W. (1975) *The Mystery of the Mind,* Princeton, N.J., Princeton University Press.

Penrose, R. (1989) *The Emperor's New Mind: Concerning Computers, Minds, and the Laws of Physics*, Oxford, Oxford University Press.

Searle, J. R., (1997) *The Mystery of Consciousness*, London, Granta Books.

Sherrington, C. (1942) *Man On His Nature*, Cambridge, Cambridge University Press.

Skarda, A. and W. J. Freeman (1987) How Brains Make Chaos in Order to Make Sense of the World, *Behavioral and Brain Sciences*, **10**, **2**, 161–195.

Slezak, P. (1983) Descartes's Diagonal Deduction, *British Journal for the Philosophy of Science*, **34**, 13–36.

Turing, A. (1950) *Computing Machinery and Intelligence*, Mind 59, 236; reprinted in D.R. Hofstadter and D.C. Dennett eds., The Mind's I, 1981, Harmondsworth: Penguin Books.

# Chapter 5

# A CELLULAR PERSPECTIVE OF NEURAL NETWORKS

Glenn E Hunt

Psychopharmacology Laboratory, Concord Hospital and
Department of Psychological Medicine, University of Sydney

How do individual brain cells combine to form neural networks that underlie all brain functions? This question can not be answered without some understanding of the building block of the network, the neuron. This chapter provides a basic overview of brain function from the cellular perspective (single neuron level) and illustrates how neurons are able to form connections with each other and continually adapt to changes in their environment.

It is important to recognize that the brain is not a fixed (hardwired) system, but a dynamic system that can respond in many ways to changes in the environment. This process known as neuronal plasticity allows the brain to learn and 're-wire' itself when damaged. I start by describing the basic elements of a neuron, how neurons communicate with each other and how this information can produce long-term changes in a neuron.

## SYNAPTIC COMMUNICATION

All of the brain's functions are carried out by nerve cells (neurons) that communicate with each other in complex networks. The 100 billion neurons that form the human brain come in many shapes and sizes. They can however be (roughly) classified as those that send sensory information (sensory neurons) to the brain to form associations with other cell types (interneurons) within the various networks, that eventually lead to control of movement (motor neurons) and hormone secretions (neuroendocrine cells).

The flow of activity from one neuron to the next can be summarized in the following path:

1) Input: a chemical (neurotransmitter) from another neuron binds to specific receptors located on dendrites much the way a key fits a lock.

2) Information from the dendrites is processed in the cell body, which decides when to generate (fire) its own action potential (electrical signal).
3) Output: the action potential travels along the axon to its terminals.
4) Synapse: a neurotransmitter is released into the gap (synapse) between the axon terminal and the dendrite of another neuron.

If one was able to follow the impulse derived from a sensory neuron through the brain to its final output, we would see that a single stimulus does not travel through the system like a car on the roadway making various turns to reach its final destination. Rather, neurons output information along their axons that terminate (synapse) on many inputs located on the dendrites (and cell bodies) of many other neurons, which in turn communicate with other sets of neurons (figure 5.1). A signal from a sensory nerve can therefore start a complex reverberation of associations that leads to further processing in the cortex.

When an action potential reaches the axon terminal, a neurotransmitter (chemical messenger substance such as norepinephrine, dopamine, glutamate, GABA, etc.) is released into the synaptic junction that diffuses to the postsynaptic dendrite and binds with specific receptors (figure 5.2). These chemical messages relay information to the dendrites that the neighboring neuron was recently activated. The transmitter can also bind to other receptors located on the axon terminal (autoreceptors) that inhibit further release (negative feedback). Other processes can augment the amount of neurotransmitter or other products such as co-transmitters or modulators to be released which further refine the message. To conserve any wastage of transmitter substance, the presynaptic terminal usually has a number of active re-uptake sites that re-cycle the transmitter back into the axon terminal where it can be re-packaged for future use.

There are many different receptor subtypes that neurotransmitters bind to on the cell surface, but they can be categorized into two main classes: 1) those linked to guanine nucleotide binding proteins (G proteins) and 2) those forming ionic channels that allow positively (sodium, calcium, potassium) and negatively charged (chloride) particles to pass through (figure 5.2). The opening of an ionic channel is very fast (<5 thousandths of a second [5 msec]) allowing for rapid communication between cells, whereas occupation of G-protein linked ligands is much slower (>30 msec). Even though a receptor may be occupied by a transmitter (or drug), this signal contributes only one piece of information relayed to the receiving neuron. Messages from several thousand receptor sites involving different receptor classes, scattered over many dendritic processes influence the cell to fire its own action potential.

## TRANSMITTERS INFLUENCE CELLULAR PROCESSES

Although most transmitters bind to their receptors for only brief periods (measured in msec), occupation of these sites can have a tremendous impact on processes contained within the cell body. The coupling of a neurotransmitter to its receptor initiates a complex cascade of events that can eventually interact with DNA contained in the cell nucleus to bring about long-term changes to cellular function.

Figure 5.3 describes this process in some detail. It shows how the signal initiated by a neurotransmitter binding to a G-protein stimulates the production of adenylate

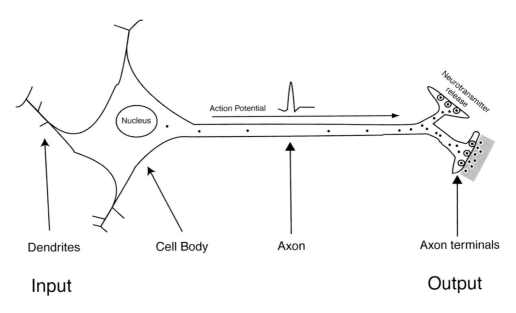

**Figure 5.1** A diagram of a neuron and its basic elements. Enlargements of the shaded synapse appear in figures 5.2 and 5.3.

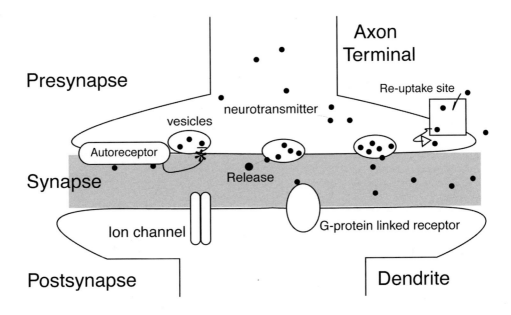

**Figure 5.2** Schematic illustration of neurotransmission. Neurotransmitters are stored in synaptic vesicles and released into the synapse (shaded area) to bind to postsynaptic receptors linked to ion channels or G proteins. Other receptors and transporters that regulate release (autoreceptor) or recycle the neurotransmitter (re-uptake site) are located on the presynapse.

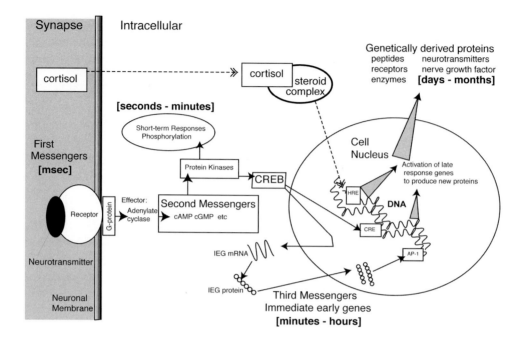

**Figure 5.3.** Diagram illustrating the intracellular dynamics of a neuron. Chemicals in the synapse (shaded area) are able to interact with DNA in the cell nucleus to bring about long-term changes to cellular function and structure. The time scale for each step is shown in brackets.

Complex network with abundant synapses formed at birth

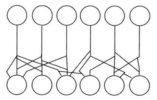

During the developmental stage many connections are eliminated if they are not co-activated

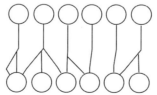

**Figure 5.4.** Diagram illustrating the retention or elimination of synaptic connections during development. The principle by which synapses are retained or eliminated is based on co-activity.

Experience molds the pattern of synaptic connections through co-activation

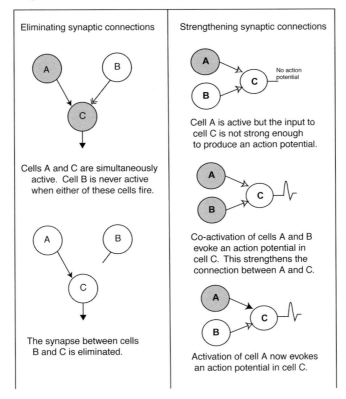

**Figure 5.5** Schematic diagram illustrating how synaptic connections are eliminated (left panel) or strengthened (right panel) through co-activation.

cyclase which in turn leads to the formation of many molecules of cAMP (cyclic adenosine monophosphate). Each molecule of cAMP activates 1 molecule of protein kinase which can phosphorylate (add phosphate) and thereby activate many copies of an enzyme that speeds up chemical reactions in the cell. Thus, the signal produced on the cell membrane is amplified at each stage within this cascade of events which allows it to affect many processes along the way. There is also a lot of cross-talk between the second messengers and their by-products from the various receptor signals received over the entire dendritic membrane. The enzymes activated by protein phosphorylation and their products are responsible for many short-term responses (like short-term memory) that last for a few seconds or minutes. However, for longer-lasting changes to occur, like long-term memory, the formation of new proteins from DNA located within the cell nucleus is required.

How do these new proteins get made in response to the information received from the cell surface? Considerable knowledge about this process has been gained by studying the molecular biology of cells across many species. Figure 5.3 shows two examples of how substances in the synaptic cleft (like cortisol, a hormone, or neurotransmitters) influence the genome in the cell nucleus and the time frame within which these effects last.

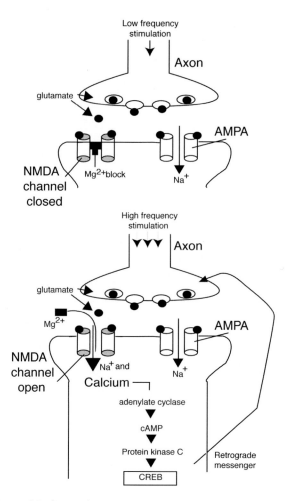

**Figure 5.6** A molecular model of strengthening synaptic connections. During low level stimulation only sodium enters the cell through the AMPA glutamate channel (upper panel). High frequency stimulation opens both glutamate channels (AMPA and NMDA) allowing calcium entry which indirectly generates a retrograde signal to the presynaptic terminal (lower panel). This results in more glutamate to be released from the presynaptic terminal the next time it receives an action potential.

The first example illustrates how cortisol (a small lipid soluble molecule) is able to pass through the outer neuron membrane and enter the intracellular space to bind to its receptor within the cytoplasm and directly interact with DNA to regulate gene expression. The second example shows how the binding of the neurotransmitter influences second messenger systems that eventually leads to the expression of immediate early genes (IEGs). The IEGs can be thought of as a 'third messenger' system transferring short-term intracellular signals into long-term changes in gene expression. Therefore the IEGs form a critical link conveying information from the plasma membrane to the cell nucleus to turn genes on and are thus important in the understanding of neuronal plasticity and repair. Some of the products regulated in this manner include peptides, enzymes and growth factors that can influence cellular function and structure for days, months and sometimes permanently.

## NEURONAL PLASTICITY

The ability of the cell to respond to novel stimuli by turning specific genes on and off allows the cell to maintain a homeostatic balance. Therefore, the introduction of novel stimuli that upsets this balance will induce changes in the cellular process leading to the manufacture of new proteins to compensate for this imbalance. For example, take the case of a drug like morphine chronically administered to a rat. The acute effects of morphine on cells located in the locus ceruleus (a region in the brainstem) produces a reduction of their firing rate. After chronic dosing, the affected cells manufacture new products (receptors, neurotransmitters and enzymes) or open and close various voltage gates to compensate for this reduction. In addition, structural changes occur to the dendritic tree. If the drug is suddenly discontinued or an antagonist like naloxone is given, then this would remove the blockage on the cell surface allowing normal information flow.

The sudden re-establishment of normal signaling would lead to an abnormally high firing pattern due to the 'up-regulation' of the firing pattern. The changed firing pattern in locus coeruleus cells has been used to explain many of the withdrawal symptoms experienced by heroin addicts. The duration of the withdrawal symptoms coincides with the time it takes the cell to adjust its firing pattern back to normal levels. This example shows how neurons can respond at the cellular level to drug effects and accounts for the increased neurotransmitter turnover, receptor number or affinity (supersensitivity) observed during chronic treatment.

Nerve damage also induces major changes in cellular signaling and function. For example, the transection of nerve fibers induces the specific production of an IEG called *c-jun* as well as products that help repair the damaged nerve. If the damage can be repaired, then *c-jun* is no longer expressed and the cell survives and begins to function normally. However, if the damage can not be repaired, then the neuron self-destructs and glial cells and macrophages scavenge the debris to remove all traces of the nerve cell. This process known as programmed cell death (apoptosis) plays an important role in neural development, discarding cells whose functions are no longer required and may also be involved in neurodegenerative states like Alzheimer's disease.

Both these examples demonstrate how novel stimuli affect the internal workings of the cell. However, changes at the cellular level can also affect surrounding neurons and are thought to be important in other long-term adaptations that affect the formation and functioning of neural networks.

## CO-ACTIVATION AND SYNAPTIC PLASTICITY

Neuronal plasticity is most active during the developmental stage when there are an excess number of connections (synapses) between cells. This over supply of synapses provides a basic wiring pattern or starting point for many neural networks to develop more precise signaling between cells. With experience, the numbers of connections are pruned back by a competitive process that involves co-activation of synaptic signaling (Figure 5.4).

This process allows for the influence of environmental factors to help shape and adapt sensory and motor processes to match different circumstances. However, if a function is lost or not used during this critical period, then the reduction in the number of synapses may be excessive and the connections that normally subserve a function may not develop (i.e. 'use it or lose it').

For example, if one eye is damaged or covered during the critical period when binocular vision is formed, the cells in the cortex responsible for stereo vision may be lost forever since they are only retained if inputs from both eyes fire in synchrony. Excess synapses are also eliminated from muscle fibers if the firing rate does not match the muscle contractions. The muscle cell rejects any synapses that were quiet while it was contracting and therefore only those axon terminals that had just been active are retained. This shows that experience acting through cellular events molds the pattern of synaptic connections in the brain (Figure 5.5). In short, 'those neurons that fire together wire together'.

Donald Hebb pioneered this idea that co-activation strengthens synaptic connections between cells which leads to functionally related groups of neurons to form cell assemblies. This basic process of strengthening connections between cells is repeated and gradually more cells are added into the assembly with further experience.

A process known as long-term potentiation (LTP) has been used to support Hebb's idea and has gained favor in learning models. LTP is observed after high frequency stimulation applied to one area, and increases the response recorded in another area, which lasts for several days. The molecular principle supporting LTP has been well studied in the hippocampus and found to involve glutamate receptors and the influx of calcium into the cell (Figure 5.6).

During low frequency stimulation the glutamate that binds to AMPA and NMDA ion channels only opens the AMPA channel because the NMDA channel is blocked by magnesium. When high frequency stimulation is applied this opens both channels because the magnesium block is dislodged allowing calcium to enter the cell. Calcium entry in turn generates a retrograde message (possibly nitric oxide) to the presynaptic terminals that were recently activated. This message leads to adaptive changes in the presynaptic terminal to allow more glutamate to be released when subsequently stimulated. This molecular model describes one of the ways synaptic signals can be strengthened between two functionally related areas.

Studies examining changes in rats placed in enriched environments (lots of toys and interesting places to explore) have demonstrated more connections (dendritic spines) in the cortex of these rats. This demonstrates an important principle that the re-wiring capabilities of the brain can occur throughout life.

The topics covered thus far have shown a variety of ways neurons can respond (adapt) to changes in environmental stimuli during the lifetime of the organism. However, this only tells half of the story because it ignores the inherited genetic influence that affects all cells in the organism. Since gene expression affects all levels of brain function, it should therefore be included as a factor in all brain models. A brief overview of genetic influences on brain function follows together with some recent findings.

## GENETIC INFLUENCES ON BRAIN FUNCTION

Humans typically have 46 chromosomes (23 pairs) made up of single molecules of deoxyribonucleic acid (DNA). DNA is the coded information contained in each cell of every living organism. This code which we inherit from our parents consists of many

DNA units called genes. From inception and during countless divisions, this genetic code determines most of our physical characteristics and tells which cells will form the various tissues in our bodies. The Human Genome Project, the largest scientific project to date, is presently mapping all the genes and their codes (amino acid sequences) contained on the human chromosomes. This information should provide valuable data to study the genetic basis of disease and to devise new treatments for people with genetic disorders.

Information must be received at specific promoter regions on the DNA before genes are activated or shut-off. Co-ordinating these signals is important, to determine why only some genes are expressed in certain cells. More genes are expressed in the brain than any other organ of the body. Recently, molecular biologists have been able to study the biological significance of genetic events by engineering foreign genes into the germ line (transgenic models) or inactivate (knockout) a gene of interest in mice. These mouse models enable neuroscientists to learn about how the brain functions when a gene is overexpressed or can no longer be expressed to assess its functional significance.

For example, several studies have shown that deleting single genes responsible for the expression of different receptor subtypes (dopamine D1, D2, D3, 5-HT1B, etc) or other important components of the cell (CREB, monoamine oxidase, protein kinase C) have profound effects on a wide range of behaviors including motor activity, sensitivity to seizures and drugs, impaired learning, enhanced aggression and changes in fear responses.

The rapid advances made in the area of molecular biology and genetics has greatly expanded the ways and means of studying brain function at various levels. No doubt future studies in this exciting area will provide new evidence to unravel the mysteries of how neurons are influenced by genes and their environment to form and maintain neural networks and their interactions at the level of the whole brain.

## REFERENCES AND SUGGESTED FURTHER READING

Alberts, B., Bray, D., Lewis, J., Raff, M., Roberts, K., and Watson, J.D. (1994) *Molecular Biology of the Cell.* New York: Garland Publishing.

Eggermont, J.J. (1998) Is there a neural code? *Neuroscience and Biobehavioral Reviews,* **22,** 355–370.

Hyman, S. E. and Nestler, E. J. (1993) *The Molecular Foundations of Psychiatry.* Washington, D. C., American Psychiatric Press.

Kandel, E.R., Schwartz, J.H., and Jessell, T.M. (1991) *Principles of Neural Science.* Norwalk, Conn: Appelton and Lange.

Kolb, B., Forgie, M., Gibb, R., Gorny, G., and Rowntree, S. (1998) Age, experience and the changing brain. *Neuroscience and Biobehavioral Reviews,* **22,** 143–159.

Nelson, R.J., and Young, K.A. (1998) Behavior in mice with targeted disruption of single genes. *Neuroscience and Biobehavioral Reviews,* **22,** 453–462.

Sakurai, Y. (1998) The search for cell assemblies in the working brain. *Behavioural Brain Research,* **91,** 1–13.

Wheal, H.V., Chen, Y., Mitchell, J., Schhner, M., Maerz, W., Wieland, H., Van Rossum, D., and Kirsch, J. (1998) Molecular mechanisms that underlie structural and functional changes at the postsynaptic membrane during synaptic plasticity. *Progress in Neurobiology,* **55,** 611–640.

**Chapter 6**

# THE BRAIN'S CHEMISTRY

Philip M Beart

Department of Pharmacology, Monash University

The previous chapter outlined the diversity of cellular mechanisms. In this chapter I focus on the brain's chemistry. All communication between cells in the brain is by chemical synaptic transmission — a unique signaling mechanism involving the targeted delivery from one neuron of minute amounts of a specific chemical on to the receptive surface of a neighboring neuron.

As outlined in the previous chapter, once sufficient action potentials reach the presynaptic nerve terminal to raise calcium ($Ca^{++}$) levels, packages (vesicles) storing specific chemicals fuse with the cellular membrane and release the chemicals into the gap (synaptic cleft) between cells, allowing the chemicals to act upon adjacent neurons. Synaptic transmission allows chemicals to interact with specific proteins in the cellular membrane (receptors) of the recipient neuron and triggers the consequent cascade of functions across second–third messengers–genes–DNA in the manner described in the previous chapter.

Networks in the brain play key roles in its signaling patterns and in determining the overall balance between inhibitory and excitatory processes across scale. Just as networks themselves are quite different in their makeup and complexity (heterogeneous in the size and shape of their individual neurons), insights into chemistry represent a further dimension of defining brain networks, since individual neurons contain different chemicals. There are a number of levels at which brain chemistry defines networks:

1) The regional distribution and the absolute concentration of a particular chemical molecule in a brain area.
2) The chemicals that a particular neuron or network contains and the manner in which it is organized in the brain.
3) Encoding a specific action, either inhibitory or excitatory.
4) Having a specific target site (receptor or enzyme) in the brain, where it acts.

Once some of these pieces of chemical information are available, not only can we refine our knowledge of existing neuronal networks, but we can begin to probe and understand ones not previously understood. From a therapeutic perspective, new synthetic chemicals (i.e. drugs) can then be employed to alter or correct a known chemical network, which may be malfunctioning or have been genetically identified as defective in a neurological or psychiatric condition.

## METABOLIC BRAIN CHEMISTRY

The brain's organizational structure (hemispheres, lobes, nuclei) is outlined in the next chapter, and I focus here on the means of communication among regions via distinct signaling patterns along defined networks. Even before we examine the chemistry of signaling in the brain, there are some simple aspects of brain chemistry we need to be aware of. The brain's metabolic and structural chemistry are unique and quite different from all other body organs. Especially, its energy consumption and its fat content. The brain's metabolic activity is exceptionally high and accounts for about 20% of the body's total oxygen consumption, and glucose is its primary source of energy. We all know that falls in blood glucose make us feel less lively and we don't think so well — in fact cerebral dysfunction occurs ranging from mild behavioral impairment to coma.

Considerable energy (the chemically relevant molecule being adenosine triphosphate [ATP]) is needed to maintain ionic balance in brain cells, since such ionic movements underlie the generation of action potentials and transfer of information between cells by chemicals. Imaginative methodology derived from the work of Sokoloff and colleagues in the USA has made it possible to examine metabolically active areas in the brain using radioactive glucose and imaging techniques (see chapter 15 on brain imaging technologies). In fact, relatively discrete brain areas and nuclei may be activated by specific tasks, drugs and pathological conditions. Since glucose is preferentially metabolized in neurons, this type of information allows the establishment of metabolic maps of brain chemistry and an understanding of physiologically active brain regions to be developed. Thus using this technology, models can be set up of how this unique chemical is important to brain networks.

Neurons have unique structural geometries and fats (such as phospholipids), cholesterol and more complex related molecules are integral parts of the membranes, receptors and sheaths covering axons that are also essential to chemical signaling. Indeed, 50% of dry weight of brain is fat compared with 5–20% of other organs. Whilst these chemicals and their biochemistry are not restricted to the brain, they are utilized rather differently in the brain because of its complex functional organization.

## CRITERIA FOR THE IDENTIFICATION OF NEUROTRANSMITTERS

There are a number of unique molecules that are specifically involved in signaling between cells in the brain; frequently they are not found in other tissues or are utilized in only a minor way. A good guiding principle is that if a chemical is used in communication in peripheral tissues, then it is likely to be a messenger molecule in the brain. However, during evolution, the brain has adapted certain simple molecules

**TABLE 6.1**
Criteria for Neurotransmitters.

1. Present in neurons with biosynthetic enzymes and stored in vesicles.
2. Release in calcium ($Ca^{2+}$) dependent manner upon depolarization.
3. Receptors exist which alter cellular excitability.
4. Specific receptor antagonists block synaptically released or externally applied substance.
5. Mechanism exists for termination of synaptic action.

for major roles in brain signaling relative to other organs. There are about 100 molecules that act as chemical neurotransmitters mediating signaling between adjacent neurons. Transmitter identification is not a simple procedure experimentally because of the brain's complex organization, the rich diversity of chemicals it contains and because it is not always possible to study individual synapses at the microscopic level. To identify whether a specific small or large molecule is a transmitter, neuroscientists have established a number of criteria that they can test experimentally (table 6.1).

Indeed, only for a dozen or so chemicals have all of these criteria been satisfied. Many other molecules are widely accepted internationally by neuroscientists as transmitters, but are still under investigation. Still others are viewed with caution and frequently referred to as putative neurotransmitters.

## SMALL AND LARGE MOLECULE TRANSMITTERS

Chemical signaling between neurons involves a rather diverse collection of chemicals, different not only in their chemical make-up, but also in their signaling mechanisms. The concept of chemical neurotransmission originated from the experiments of Otto Loewi in Germany (1921), who showed that stimulation of one frog heart produced chemical activity which could be used to activate the heart of another frog. The phenomenon of chemical neurotransmission was first explored in peripheral nerves, particularly the nerve-muscle junction, before it was shown to apply to the brain. Although some 100 specific molecules are capable of transferring a signal between nerve cells, there are a number of discrete classes (essentially classified as small or large) chemical neurotransmitters.

Small molecule transmitters are often referred to as the 'classical' transmitters: they have molecular weights <200. These simple chemicals were the first molecules identified as participating in the transmission of information between neurons and are thus the best studied class of neurotransmitter. In fact, most of our basic principles of chemical synaptic neurotransmission have arisen from studies of these small molecules. It is a fair generalization that these guiding principles are conserved across synapses that differ morphologically and in the identity of their transmitter, and across all animal species examined.

The first discovered subgroup of small molecules were the biogenic amines. Acetylcholine was identified in the 1920s: its actions have been thoroughly investigated at the nerve–muscle junction. Other biogenic amines include the catecholamines (norepinephrine, epinephrine and dopamine), which were widely studied by Swedish scientists in the 1950s–60s using fluorescent histochemistry to provide the first

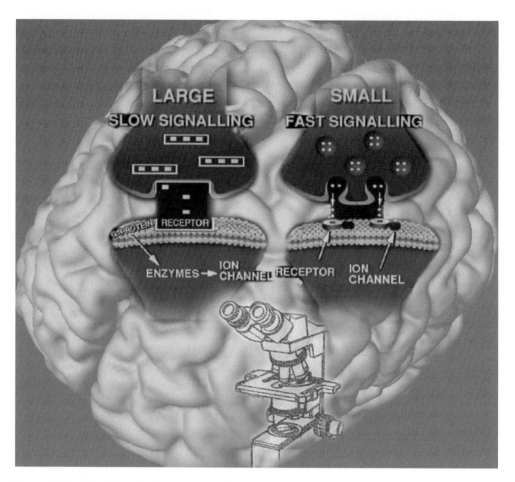

**Figure 6.1** Small and large brain neurotransmitters.

chemical brain maps of these remarkable, long-process neurons (see the brain's chemicoarchitecture in the next chapter).

The other major subgroup of small molecule neurotransmitters are the amino acids. These simple molecules are integrally involved in cellular biochemistry and represent the building blocks of proteins. Because of their roles in intermediary metabolism and widespread distribution, it proved difficult to establish their specific roles in neurotransmission and to fulfill the criteria listed in table 6.1. They have really only been studied since the 1960s and are now considered to be quantitatively the major group of neurotransmitters in the brain. L-glutamate is the major transmitter of excitatory information in brain and may be involved at approximately 50% of all synapses. Simple amino acids also mediate inhibition; GABA (4-aminobutyric acid) is the major inhibitory neurotransmitter and has been estimated to be used at some 25% of all synapses in the brain. Glycine, the simplest amino acid, is also an inhibitory neurotransmitter, particularly in the spinal cord and brainstem (pons and medulla oblongata). A number of other simple molecules also appear to be involved in chemical signaling between neurons (see table 6.2).

**TABLE 6.2**
Chemical Neurotransmitters.

SMALL MOLECULE TRANSMITTERS:
Biogenic amines:                          acetylcholine, norepinephrine, epinephrine, dopamine,
                                          serotonin
Amino acids:                              L-glutamate, L-aspartate, GABA, glycine
Miscellaneous:                            adenosine, nitric oxide

LARGE MOLECULE TRANSMITTERS:
Neuropeptides:                            (many amino acids)
Thyrotropin releasing hormone:            (3 amino acids)
Methionine:                               (5 amino acids)
Enkephalin:                               (5 amino acids)
Angiotensin II:                           (8 amino acids)
Cholecystokinin:                          (8 amino acids)
Substance P:                              (11 amino acids)
Neuropeptide Y:                           (36 amino acids)

The other major group of neurotransmitters is the large molecules. These are generally neuropeptides and are assemblies of some 3–50 individual amino acids, with molecular weights ranging from 200–5000. Neuropeptides have been recognized for many years to be used in endocrine and intestinal systems and to play messenger roles upon release into the blood and gut, but are now known to mediate synaptic transmission in the brain. Many peptides playing important physiological roles in peripheral body organs have subsequently been demonstrated to be neurotransmitters. More than 50 such neuropeptides have been identified and new discoveries are still being made. Neuroactive peptides have been particularly difficult to establish as neurotransmitters in the brain, because of their very low concentration (some 3-fold less than amino acids and amines) and because of the lack of suitable pharmacological tools. Most peptides only partially fulfill the list of transmitter criteria (table 6.1), but there are continuing advances in this very active area of research.

One such peptide is substance P (originally detected in a powder of a tissue extract) first discovered in 1931, and not actually identified until 40 years later. With the purification of its receptors and the synthesis of biologically active receptor antagonists, substance P is generally accepted by neuroscientists to be a neurotransmitter. Two other peptides, leucine- and methionine-enkephalin, which act upon morphine-like (opiate) receptors, were isolated from the brain in the 1970s and are known to be involved in the brain's sensation of pain. Strictly speaking many peptides should be classified as 'neuroactive', but scientists now believe that in excess of 30 peptides are neurotransmitters in numerous regions of the brain. A number of peptide families are recognized and these are likely to be involved in physiological functions as diverse as blood pressure regulation, food intake, memory and hormone secretion.

## SLOW AND FAST SYNAPTIC TRANSMISSION

The synaptic mechanisms and particularly the time-course of the signaling processes are quite different for small and large molecule transmitters. Whilst synaptic transmission involving small molecule transmitters has been thoroughly analyzed for at least 30 years, only comparatively recently with the advent of modern sophisticated techniques (especially those of molecular biology) have many of the fine details been elucidated. Signaling mediated by neuropeptides has been studied for a shorter period of time and is less well understood.

Small molecule transmitters are synthesized in the presynaptic nerve terminal, where the appropriate precursor molecule and the enzyme (a specialized protein) necessary to achieve the synthesis can also be found. In fact, this enzyme is yet another chemical signature indicating the specific involvement of a discrete transmitter. For example, choline acetyltransferase with acetylcholine, glutamate decarboxylase with GABA, and tyrosine hydroxylase with norepinephrine or dopamine. Antibodies against specific enzymes have been employed routinely to localize specific transmitter systems. The newly synthesized transmitter is assembled into specialized packets termed 'vesicles', which are generally small.

During propagation of the action potential, the vesicles move towards and subsequently fuse with the presynaptic membrane to empty their contents at small, discrete release zones in close proximity to the postsynaptic membrane of a connected neuron. The interaction of the neurotransmitter with a specific receptor in the postsynaptic membrane leads to a rapid electrical change, either a depolarization (excitatory) or a hyperpolarization (inhibitory). Often these receptors are directly linked to an ion channel and the neurotransmitter action is so fast as to take only a few milliseconds. Molecular biological techniques have allowed the sequencing and cloning of these receptor-channel complexes and they represent a family of proteins sharing some structural similarities, including binding sites for modulatory molecules.

Once the transmitter has exerted its postsynaptic action, the event is efficiently terminated by a transport process, which carries the released transmitter back into the presynaptic nerve terminal. This transport process is sodium ($Na^+$) and energy dependent, and serves the purpose of recycling transmitter molecules. Many parts of the synaptic machinery can be modulated by drugs, and the pharmaceutical industry has spent considerable time and money attempting to direct chemical molecules at discrete synaptic targets. For example: drugs that block the serotonin transporter are useful in the management of depression; dopamine receptor antagonists are widely used in the treatment of schizophrenia; Valium, a benzodiazepine prescribed for anxiety, binds to a specific site in a receptor for GABA.

The neuropeptide, large molecule transmitters share some common features with signaling mediated by small molecules, but there are a number of important differences. Firstly, neuropeptides are not synthesized in the presynaptic nerve terminal, but their synthesis involves breakdown of a much larger polypeptide to form the neuroactive peptide species in the cell body. This precursor polypeptide is packaged into vesicles and actively transported down the axon to the nerve terminal, where further local synthesis may occur. Overall, these mechanisms are quite different from those of the 'classical' small molecule neurotransmitters, which are synthesized in the presynaptic nerve terminal. A second unique feature of peptidergic neurons is that they tend to release their peptides from vesicles at less specialized release zones. Also, there is evidence of the release process occurring at higher frequencies of stimulation than required to release small molecule transmitters.

Thirdly, the time-course of the chemical signaling mediated by neuropeptides is often quite slow, being of the order of seconds, minutes or even hours. Such slow synaptic transmission reflects the type of postsynaptic receptor involved. These receptors rather than being directly linked to an ion channel, instead activate specific enzymes (as described in the previous chapter). Fourthly, the postsynaptic action of neuropeptides is not terminated by active transport from the synaptic cleft, but rather by the slower process of enzymatic cleavage.

## FUNCTIONAL ORGANIZATION OF NEUROTRANSMITTERS

The patterns of organization of neurons have always excited neuroscientists who believed that the connectivity details might shed light on the nature of integrated function. Countless studies using lesions and neuronal tracing techniques, in conjunction with the localization of various chemical signatures (chemicals, enzymes, receptors, transporters) by histochemistry and the use of antibodies, have provided neuroscientists with a good idea of how neurons are topographically distributed.

There are some guiding principles on how neurons and their specific chemicals are organized. Simple, fast acting transmitters are generally contained in neurons with long axonal processes which distribute their synapses widely. Neurons containing these biogenic amines are frequently tightly clustered and send long-axoned processes to many brain regions. For example, dopamine neurons with their cell bodies in the substantia nigra project to and release their transmitter in the neostriatum. This population of neurons is known to be important in motor activity and degenerates in Parkinson's disease. Neurons containing acetylcholine are similarly organized and one interesting cholinergic pathway arises in the basal forebrain, activates the cortex and hippocampus, is involved in learning and memory and is defective in senile dementia of the Alzheimer's type. Other rapidly acting transmitters, GABA and L-glutamate, are also present in neurons with long processes, although they tend to be more widely distributed.

Neuropeptides, by comparison, are involved in slower synaptic transmission and are often localized in a small type of neuron with short processes called a local circuit interneuron. They generally modulate the activity of the long process neurons in brain. These basic patterns of neuron interaction constitute the beginnings of functional networks.

## DIVERSITY AND INTEGRATION OF NEUROTRANSMISSION

Chemical synaptic transmission is often in practice somewhat more complex than the description provided here, which serves to illustrate the basic characteristics of chemical signaling between neurons. With recent advances we now know some small molecules are both fast and slow transmitters (e.g. norepinephrine, L-glutamate, GABA, serotonin). Thus L-glutamate, the major excitatory transmitter, exerts its actions at both ion channel-linked ('ionotropic'; fast) and second messenger-linked ('metabotropic'; slow) receptors. These receptors are structurally quite different with the metabotropic receptors being coupled to an effector G-protein.

Signaling diversity provided by different time courses and by the existence of multiple receptors for a single transmitter allows for differential influences on neuronal activity and for fine-tuning of brain networks (and this diversity should be incorporated into brain model developments).

A further complexity, which is not well understood at the present time, is the phenomenon of *co-transmission*. There are now many examples of neurons containing multiple transmitters, particularly a small molecule transmitter and a neuropeptide. Thus there appears to be co-localization within a single population of neurons and co-release of multiple transmitters. However, for many decades a widely accepted guiding principle arising from the work of the British physiologist Sir Henry Dale was that each neuron released only a single neurotransmitter. We still do not know if co-transmitters are stored in the same or different populations of vesicles, or whether they are released separately or together. It is quite likely that they are released differentially as experiments in peripheral nerves suggest that more rigorous stimulation is required to release a neuropeptide from nerves containing co-transmitters. Functionally the phenomenon of co-transmission provides yet further mechanisms for the subtle modulation of the activity of a network; the transmitters could act both slowly and rapidly to influence any neuron, they could act on different targets and they could modulate each others actions.

Any one neuron will receive hundreds (perhaps thousands) of inputs from other neurons. Each input may affect the neuron's membrane potential and these are then integrated to produce a net membrane potential, which is influenced by the frequency of firing of the inputs, their sign and summation (excitatory/depolarization or inhibitory/hyperpolarization), the spatial pattern and summation (inputs on to a cell body are more effective than those onto dendrites), the temporal pattern and whether inputs have a slow or fast time-course.

Whilst many brain areas, in isolation, can be considered to be overall inhibitory or excitatory based on their neuronal networks and major output neuronal population, it is difficult to integrate such units into a total model of activity. One approach is to consider the circuitry of a particular brain region, and evaluate the overall balance of excitatory : inhibitory activity (and factors modulating this balance). For example, the cerebellum is considered to be mainly inhibitory in function. The cortex receives mainly excitatory information and after local processing the output is mainly glutamatergic excitatory. These neurons are modulated by local circuit interneurons which use inhibitory GABA as their neurotransmitter. Such a simple model really only takes account of pathways using the brain's major transmitters. The quantitatively minor inputs by neurons containing biogenic amines also need to be considered. Models always have their limitations particularly for such a complex network. No model can be divorced from the chemistry underlying synaptic transmission, and ultimately no model can ignore the dynamical chemical interactions underlying all brain functions.

## CHEMISTRY OF NEUROLOGICAL AND PSYCHIATRIC CONDITIONS

The functional 'chemoarchitecture' of the brain is still largely unknown (i.e. precisely what functional role specific neurotransmitters have). A number of speculative associations have been prevalent in the scientific literature including: linking acetylcholine with memory; dopamine with reward; endorphin with pain; norepinephrine with

**TABLE 6.3**
Chemistry of Neurological and Psychiatric Conditions.

| Disorder | Neuropathology | Chemistry | Treatment |
| --- | --- | --- | --- |
| Parkinson's disease | Basal ganglia | 90% loss dopamine | Dopamine precursor (L-dopa and dopamine mimetics) |
| Huntington's chorea | Basal ganglia | 50% loss GABA | Elevate GABA level |
| Alzheimer's disease | Frontal and temporal lobes, basal forebrain | Loss acetylcholine, serotonin and norepinephrine | Stimulate cholinergic neurons, cognitive enhancers, smart drugs |
| Schizophrenia | Temporal and frontal lobe, amygdala, hippocampus | Elevation of dopamine and serotonin receptors | Dopamine and serotonin receptor antagonists |
| Depression | Frontal and temporal lobes | Norepinephrine and serotonin receptors altered | Norepinephrine and serotonin transporter inhibitors |
| Anxiety | Not expected, transient | Essentially no information | Valium, barbiturates, alcohol, antidepressant drugs |

'fight and flight' networks and memory; histamine with arousal; melatonin with our biological circadian clock; serotonin with aggression and mood.

Chemical insights about the details of synaptic function have been more clearly determined. It seemed only logical for neurochemists to look for chemical deficits in human postmortem brain tissue. Studies of this type were instrumental in understanding Parkinson's disease, where an 80–90% loss of dopamine in the basal ganglia was found to be associated with the degeneration of dopamine neurons with their cell bodies in the substantia nigra and sending long axoned processes to the neostriatum. Similar investigations in Huntington's chorea revealed a 50% loss of GABA in the neostriatum associated with the death of projection neurons and interneurons utilizing GABA. Information on the specific deficits in a chemical transmitter in a neurological or psychiatric disorder can allow the design and targeting of a drug to rectify that imbalance and ameliorate the clinical symptoms (see examples in table 6.3, and in figure 1.17 in The Big Picture).

## CONCLUSIONS AND THE FUTURE

Studies of brain chemistry have led to important advances in our understanding of how single neurons function individually and how they underlie network activity. Chemical neuroanatomy as a science (where neurons containing a specific transmitter are found in the brain and where their receptors are localized) has advanced greatly in the last decade. Molecular biology has also had a tremendous impact on brain chemistry, particularly with regard to receptors which have been cloned, sequenced and modeled (including which segments bind drugs and activate specific intracellular messengers). Such information means that new drugs selective for a subunit of a receptor can be developed — if we know that a subunit is found only in a precise area of brain then drug targeting may be possible.

Molecular biology, and in particular its application to the genetics of neurological diseases, will provide new information on the chromosomes involved and eventually genetic testing of individuals will assist in identifying those individuals at risk. Techniques for brain repair have already been successful in Parkinsonian patients although they are clearly in their infancy — we can expect eventually that viable neurons expressing a particular chemical transmitter or receptor, or growth factor, or an enzyme to synthesize a specific transmitter, will be able to be transferred successfully into patients with a neurodegenerative disease. Such methodology is not likely to be widely available over the next decade.

In the meantime we can expect major advances in the management of neurological conditions using drugs targeting neuropeptides. These mediators of slow synaptic transmission are suitable for drug regimes and advances are being made on how these peptidergic drugs need to be delivered to the brain for patients to receive clinical benefits. Appropriate combinations of drugs for specific disorders are being determined and explored in an increasingly fundamental fashion, in conjunction with more sophisticated models of brain function.

Chemistry is fundamental to synaptic transmission, and essential to our understanding of the integrative mechanisms of human brain function.

## REFERENCES AND SUGGESTED FURTHER READING

Albin, R. L., Young, A. B. and Penney, J. B. (1989) The functional anatomy of basal ganglia disorders, *Trends in Neurosciences,* **12**, 366–375.

Bachy-Rita, P. (1994) The brain beyond the synapse: a review, *Neuroreport* **5**, 1553–1557.

Bean, A. J., Zhang, X. and Hokfelt, T. (1994) Peptide secretion: what do we know? *FASEB Journal,* **8**, 630–638.

Birnbaumer, L. (1990) G proteins in signal transduction, *Annual Review of Pharmacology and Toxicology,* **30**, 675–705.

Catteral, W. A. (1988) Structure and function of voltage sensitive ion channels, *Science,* **242**, 50–61.

Changeux, J. P. (1993) Chemical signaling in brain, *Scientific American,* **269**, 30–37.

Cooper, J. R., Bloom, F. E. and Roth, R. H. (1991) *The Biochemical Basis of Neuropharmacology*, Oxford University Press, New York, 6th ed.

de Felipe J. and Farinas, I. (1992) The pyramidal neuron of the cerebral cortex: morphological and chemical characteristics of the synaptic inputs. *Progress in Neurobiology,* **39**, 563–607.

Francis, P. T., Pangalos, M. N. and Bowen, D. M. (1992) Animal and drug modeling for Alzheimer synaptic pathology, *Progress in Neurobiology,* **39**, 563–607.

Fried, G. (1995) Synaptic vesicles and release: new insights at the molecular level, *Acta Physiologica Scandinavica,* **154**, 1–15.

Hall, Z. W. (1992) *An Introduction to Molecular Neurobiology*, Sinauer Associates, Sunderland.

Katz, B. (1966) *Nerve, Muscle and Synapse*, McGraw-Hill, New York.

McGeer, P. L., Eccles, J. C. and McGeer, E. G. (1987) *Molecular Neurobiology of Mammalian Brain*, Plenum Press, New York.

Nicoll, R. A. (1988) Coupling of neurotransmitter receptors to ion channels in the brain, *Science,* **241**, 545–551.

Sedvall, G., Farde, L., Persson, A. and Wiesel, F. A. (1986) Imaging of receptors in the living human brain, *Archives of General Psychiatry,* **43,** 995–1006.

Sollner, T. and Rothman, J. E. (1994) Neurotransmission: harnessing fusion machinery at the synapse, *Trends in Neurosciences,* **17,** 344–348.

Shepherd, G. M. (1974) *The Synaptic Organization of the Brain,* Oxford University Press, New York.

Strange, P. G. (1992) *Brain Biochemistry and Brain Disorders,* Oxford University Press, Oxford.

Verhage, N., Ghijsen, W. E. J. M. and da Silva, F. H. L. (1994) Presynaptic plasticity: the regulation of $Ca^{2+}$- dependent transmitter release, *Progress in Neurobiology,* **42,** 539–574.

Young, S. and Concar, D. (1992) These cells were made for learning, *New Scientist,* **136** (supplement 3)

# Chapter 7

# THE BRAIN'S ANATOMY

Ken Ashwell[1], Elizabeth Tancred[1] and George Paxinos[2]

The School of Anatomy[1] and Psychology[2], The University of New South Wales

The nervous system, and in particular the brain, is concerned with perceiving changes in the environment, determining appropriate responses to those changes, and carrying out optimal adaptive behavior. On this simple level, the nervous system can be considered to consist of:

1)  *Input* sensory systems for receiving and analyzing information about the environment,
2)  *Output* systems which include voluntary motor, speech and autonomic components for either acting upon the external environment or altering conditions within the body,
3)  *Association* systems in between 1 and 2, for combining information from several senses and deciding on the appropriate course of action in the light of past experience.

See figures 1.2–1.4 in The Big Picture and the figures in this chapter for a visual distilation of these networks.

## BASIC STRUCTURAL ELEMENTS

The nervous system can be divided into two basic anatomical regions. Table 7.1 shows that these are the central nervous system (CNS) which consists of the brain and spinal cord, and the peripheral nervous system (PNS) which consists of all the other nervous tissue in the body. Overlapping with these two basic anatomical subdivisions is the autonomic nervous system (ANS) which is concerned with control of the organs (in the chest, stomach and pelvis) as well as some elements in the skin (e.g. sweat glands). Most of the ANS lies within the peripheral nervous system, while some of its elements are found within the CNS.

TABLE 7.1

The Nervous System

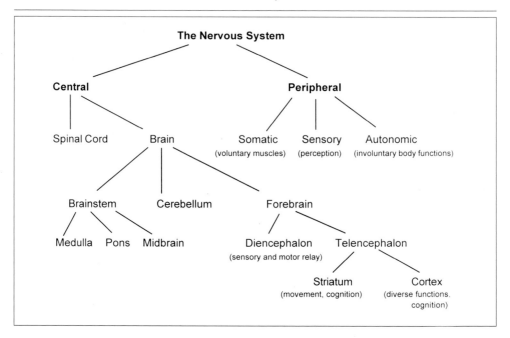

Cell types within the central nervous system fall into three groups. These are neurons, glia and cells of the vasculature. Neurons are considered to be responsible for the transmission and analysis of all electricochemical information within the nervous system. In chapter five, an illustration of the basic features of a typical neuron was provided. Found within the cell body of the neuron is the nucleus, where genetic information is stored and accessed for control of metabolic activity. The nucleolus, in turn, lies within the nucleus. Surrounding the nucleus is the cytoplasm, which includes mitochondria (responsible for producing energy in a usable form), rough endoplasmic reticulum (for the production of proteins) and the Golgi apparatus (for packaging of proteins and other material prior to transport throughout the cell). Extending from the soma are two types of processes: dendrites and axons. Dendrites are often multiple, are shorter than axons, and convey inputs of electrical activation passively from their tips towards the soma. Axons may be up to a meter in length, possess a myelin (fat) sheath and convey electrical activation away from the soma by an active form of propagation known as the action potential. The direction of flow of electrical information within the neuron is from the dendritic tips towards the soma and thence along the axon to its end (see chapter five for functional details). Note that when axons are grouped together in the CNS they are known as a tract or column. An aggregation of neurons in the brain is known as a brain nucleus, a term which is unfortunately the same as that used for a cell's nucleus.

The end branches of the axon are known as telodendria and these in turn end in rounded structures known as synaptic boutons. These boutons make specialized contacts known as synapses with dendrites, axons or the cell bodies of other neurons. Chemicals known as neurotransmitters are produced by the presynaptic neuron, packaged in structures known as synaptic vesicles. When an action potential reaches the bouton, these vesicles fuse with the cell membrane at the synapse, releasing the neurotransmitter into the synaptic cleft between the two neurons. The neurotransmitters act on receptors situated on the postsynaptic membrane to alter its electrical properties, thus influencing the electrical or biochemical activity of the postsynaptic neuron (see chapter six for further biochemical details).

The main glial types are oligodendroglia, which form fat sheaths around neurons, astrocytes, responsible for control of ions in the CNS and microglia, which form part of the immune system. While glia are usually given only secondary consideration, they are of major importance in CNS function. For example, recent evidence suggests a role for astrocytes in some types of memory.

## EMBRYOLOGICAL DEVELOPMENT OF THE CNS

The brain and spinal cord in the embryo develop from a tubular structure, the neural tube. Three dilated vesicles appear in the brain end of the neural tube at about the end of the fourth week of gestation. These are known as the forebrain, midbrain and hindbrain vesicles (see figure 7.1). Two outgrowths, known as telencephalic vesicles, appear from the sides of the forebrain vesicle, the remainder of which will become the diencephalon. The hindbrain vesicle divides into the metencephalon at the front (which gives rise to the pons and cerebellum) and the myelencephalon which gives rise to the medulla which links with the spinal cord. In the adult, the midbrain, pons and medulla are collectively called the brainstem (see figure 7.5).

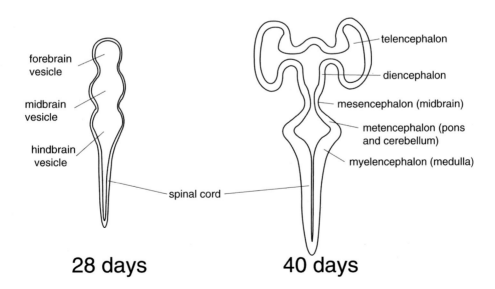

**Figure 7.1** Embryological development of the CNS.

## A) human brain - lateral view

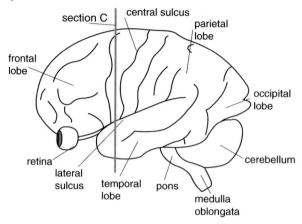

## B) human brain - medial view

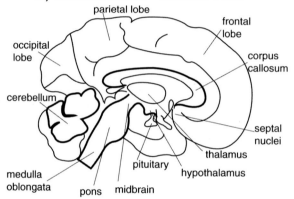

## C) human brain - cross section

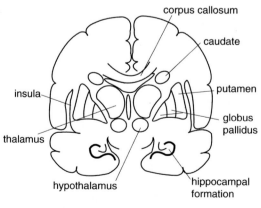

**Figure** 7.2 Gross features of the adult human brain as seen in a lateral view (A), a medial view (B) and a representative cross section (C).

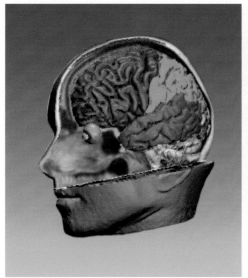

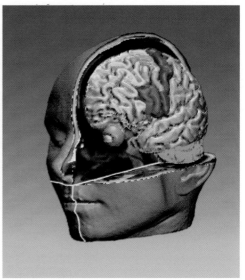

**Figure 7.3** (above left) Shows the brain's outer cortical convolutions and four lobes (Frontal in red, Parietal in green, Temporal in purple and Occipital in yellow).

**Figure 7.4** (above right) The primary motor strip is in red, the motor write area is in grey, the motor speech area (Broca) in purple, the secondary motor area in yellow and the frontal eye field is in green. *Courtesy of Voxel — Man, Brain and Skull (Karl Heinz Höhne and Springer electronic media).*

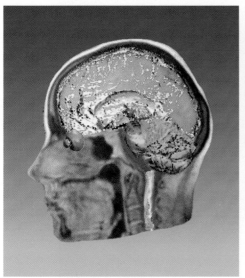

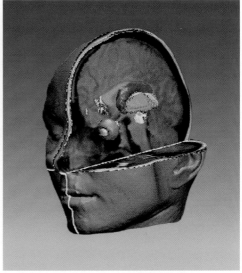

**Figure 7.5** (above left) Structures of the midbrain (green), and hindbrain consisting of pons (yellow), cerebellum (light blue) and medulla (purple).

**Figure 7.6** (above right) Structures of the basal ganglia, including the caudate (purple), putamen (green) and the thalamus in yellow. *Courtesy of Voxel — Man, Brain and Skull (Karl Heinz Höhne and Springer electronic media).*

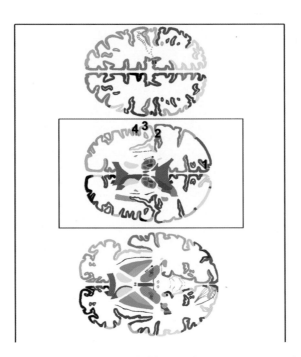

**Figure 7.7** Subdivision of the cortex on the basis of fibre and neuron stains according to Brodmann (52 separate regions). Transverse slices (front to the left) with primary occipital cortex (1), superior temporal (2), the post-central (3) and 4 (precentral) areas numbered. *Courtesy of The Electronic Clinical Brian Atlas (Nowinski, Bryan and Raghavan — Thieme, New York and Stuttgart).*

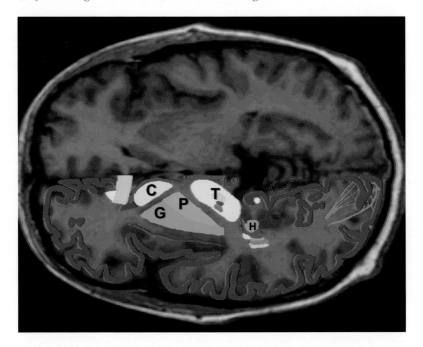

**Figure 7.8** Transverse MRI section from Talairach and Tournaux (1988) with structures color coded in half the image (C: caudate; G: globus pallidus; P: putamen; T: thalamus; H: hippocampus). *Courtesy of The Electronic Clinical Brain Atlas (Nowinski, Bryan and Raghavan — Thieme, New York and Stuttgart).*

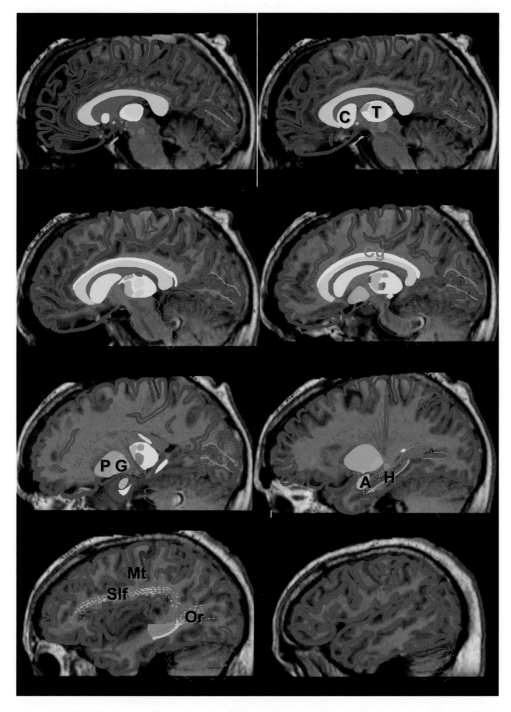

**Figure 7.9** Sagittal sections (C: caudate; T: thalamus; Cg: cingulate; P: putamen; G: globus pallidus; A: amydgala; H: hippocampus; Mt: motor tract; Slf: superior longitudinal fasciculus; Or: optic radiations). *Courtesy of The Electronic Clinical Brain Atlas (Nowinski, Bryan and Raghavan — Thieme, New York and Stuttgart).*

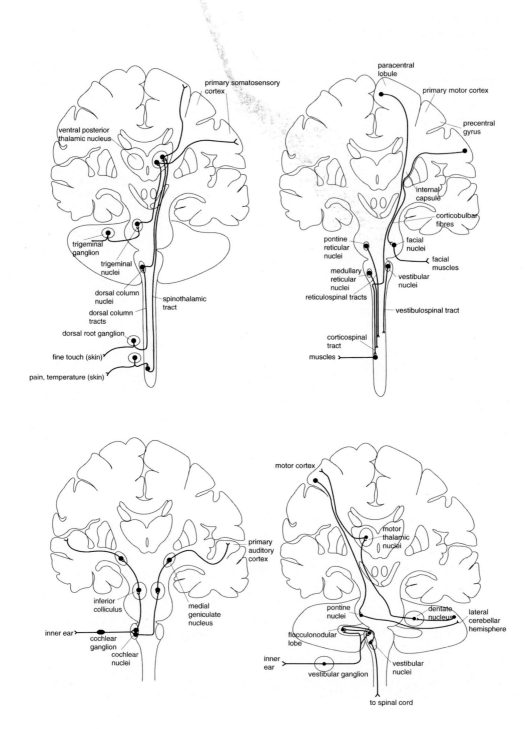

**Figure 7.10** Summary diagram of the somatosensory pathways, the auditory system, major descending motor pathways to the brainstem/spinal cord and major connections of the cerebellum.

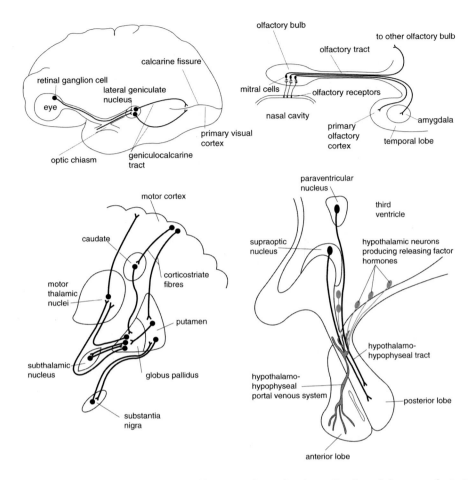

**Figure 7.11** The human visual system, olfactory pathway, basal ganglia, hypothalamus and pituitary connections.

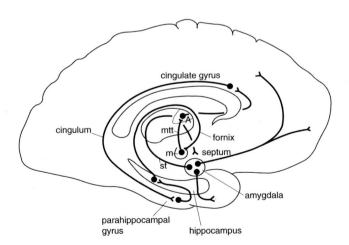

**Figure 7.12** Major components and fibre tracts of the limbic system (A: anterior nucleus of the thalamus; mtt: mammillothalamic tract; m: mammillary body; st: stria terminalis).

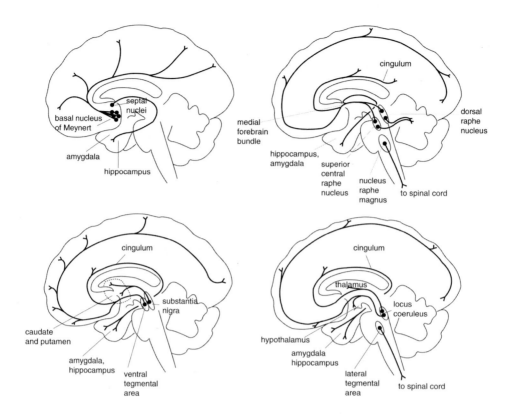

**Figure 7.13** Major cholinergic (left top), serotonergic (right top), dopaminergic (left bottom) and norepinephric pathways (right bottom) in the brain.

## ADULT BRAIN STRUCTURE

The adult anatomy of the brain can be considered according to the developmental plan outlined in table 7.1. Figures 1.2–1.4 in The Big Picture and the figures in this chapter show many of the details of the brain's architecture outlined in the text. An average figure for adult human brain volume is approximately 1,400 cm³ and it has been estimated that there are approximately $10^{11}$ neurons in the human brain (Nolte, 1993).

The most prominent feature of the telencephalon of the adult is the cerebral cortex, which covers most of the external surface of the brain, thus obscuring deeper parts of the telencephalon such as the basal ganglia and hippocampus. The cerebral cortex is divided into 5 lobes. Four of these (the frontal, parietal, temporal and occipital lobes) are visible on the lateral (outer) and medial (inner) surfaces of the cerebral hemispheres, while the 5th (the insula) is hidden in the depths of a fissure (the lateral fissure), which separates the frontal from the temporal lobes. The cortical surface is thrown into folds known as gyri separated by grooves which are known as sulci. This allows for the accommodation of a very large surface area of cerebral cortex within the skull. In fact, approximately two-thirds of the cortical surface is hidden in the depths of the sulci. Beneath the cortex lies a large mass of white matter, containing all the connections between cortical areas and deeper structures. A large fibre bundle, known as the corpus callosum can be seen on the medial surface of the forebrain. It contains about 200 million axons and provides a means for information to be transferred between the two hemispheres. The basal ganglia consist of a collection of brain nuclei at the base of each cerebral hemisphere. These nuclei include the caudate, putamen and globus pallidus (see figures in The Big Picture section and in this chapter). The septal nuclei occupy a small but important area of the brain immediately below the front end of the corpus callosum near the midline.

The center of each hemisphere of the telencephalon contains a large cavity known as the lateral ventricle, which is filled with cerebrospinal fluid (CSF). The lateral ventricles communicate through the interventricular foramina with the third ventricle, a midline cavity between the two halves of the diencephalon.

The diencephalon consists principally of the thalamus and hypothalamus. The thalamus lies in an intermediate position between the telencephalon and the brainstem. As such, its function is as something of a central relay station conveying information from the senses, spinal cord and brainstem to the cerebral cortex (see figure 1.2 in The Big Picture). It also is concerned with control of motor function, because several of its component nuclei are part of feedback loops for motor co-ordination (figures 1.3 and the Alexander model in The Big Picture). The hypothalamus, as its name implies, lies below the thalamus. It is anatomically and functionally connected with the pituitary, the brain's master organ for release of hormones for control of reproductive and metabolic function (the hypothalamus is also interconnected to the brainstem autonomic centers).

The midbrain contains relay and reflex centers for visual and auditory function in its dorsal (upper) part, and fibre pathways for motor and somatosensory function (pain, temperature and touch) passing through its ventral (lower) part. Several nuclei in this region contain neurons projecting to the forebrain and releasing neurochemical substances (neurotransmitters) such as dopamine and serotonin (see chapter 6 and figure 7.13).

The pons and medulla contain many ascending and descending fibers joining the spinal cord and cerebellum with higher parts of the CNS. Twelve cranial nerve nuclei involved in control of the muscles and glands of the head and neck and the reception of sensory information are located here (see table 7.2). The brainstem also contains groups of nuclei which are concerned with the control of respiratory and cardiovascular function. These nuclei form a central core of grey matter in the brainstem known as the reticular formation. Some nuclei in this region have neurons which produce the neurotransmitters serotonin and norepinephrine and project to diverse regions of the brain and spinal cord.

Attached to the pons posteriorly is the cerebellum, which is involved in maintaining the smooth integration of muscle activation in motor activities.

## INPUT SENSORY SYSTEMS

Specialized networks of sight, hearing and touch have a similar organizational pattern in the brain, with inputs at receptors travelling via the thalamus to primary areas of the cortex where their fundamental features are integrated in the secondary and association cortices (possibly in a hierarchical manner — see figure 1.2 and the Luria model in The Big Picture section).

TABLE 7.2
The Cranial Nerves.

| Number | Name | Function | Attached to |
|---|---|---|---|
| 1 | olfactory | smell | telencephalon |
| 2 | optic | vision | diencephalon |
| 3 | oculomotor | eye movements, pupil size and focusing | midbrain |
| 4 | trochlear | eye movements | midbrain |
| 5 | trigeminal | touch, pain from face chewing | pons |
| 6 | abducens | eye movements | pons/medulla junction |
| 7 | facial | facial movements, taste salivary and tear glands | pons/medulla junction |
| 8 | vestibulocochlear | hearing, balance | pons/medulla junction |
| 9 | glossopharyngeal | taste, salivary glands | medulla |
| 10 | vagus | taste, autonomic function speech | medulla |
| 11 | accessory | neck movements | medulla |
| 12 | hypoglossal | tongue movements | medulla |

## The Somatosensory Pathway

This system is concerned with the perception of pain, touch and temperature from the body surface, mouth, nose and throat. Receptors located in the skin are innervated by nerve processes from dorsal root ganglion cells or cranial ganglion cells situated along the spinal cord or near the brainstem, respectively.

The large central processes of some dorsal root ganglion cells enter the white matter of the spinal cord and ascend as large tracts known as the dorsal columns (concerned with fine touch, joint position and vibration) to end upon neurons in two nuclei of the medulla (the gracile nucleus for the lower limb and the cuneate nucleus for the upper limb). These two nuclei, in turn, project to the ventral posterior thalamic nucleus (V in figure 1.2 of The Big Picture and figure 7.10).

Other smaller dorsal root ganglion cell fibers (concerned with pain, temperature and simple touch) make synaptic contact with neurons in the dorsal horn of the spinal cord. These neurons, in turn, send axons in a fibre bundle known as the spinothalamic tract to end also upon neurons in the ventral posterior thalamic nucleus.

Cranial nerve ganglion cells send axons into the brainstem to end upon neurons in the trigeminal sensory nuclei. These neurons, in turn, send axons to terminate upon neurons in the ventral posterior thalamic nucleus. This latter nucleus is functionally subdivided, such that the lateral part is concerned with body sensation, while the medial part is concerned with the head.

Neurons within the ventral posterior thalamic nucleus send axons to a part of the cortex known as the primary somatosensory cortex, which occupies the postcentral gyrus and the posterior parts of the paracentral lobule. This region is somatotopically organized, such that the head is represented laterally and the upper and lower limbs progressively more medially.

## The Visual Pathway

Information about the visual world in the form of light falls upon the light sensitive photoreceptors (rods and cones) of the retina. After some processing of this information in the retina, retinal ganglion cells carry the information to the visual thalamic nucleus, the lateral geniculate nucleus (L in figure 1.2 of The Big Picture), which in turn projects to a part of the cerebral cortex known as the primary visual cortex (figure 7.11). This is located in the occipital lobe along the walls of the deep calcarine fissure. The primary visual cortex is visuotopically organized, such that central parts of the visual field are represented in a large portion of the cortex posteriorly, while more peripheral parts are represented in a smaller area anteriorly.

## The Auditory Pathway

The sensory receptors for hearing are the hair cells located in the inner ear. Auditory information is conveyed to the CNS by the neurons of the cochlear ganglion. The central processes of these cells end upon neurons in the cochlear nuclei of the brainstem. These cochlear neurons in turn project via a fibre bundle known as the lateral lemniscus to the inferior colliculus of the midbrain, which relays the auditory information to the appropriate nucleus of the thalamus, the medial geniculate nucleus (M in figure 1.2 of The Big Picture section). Cortical processing of auditory information occurs in the primary auditory cortex, which receives axons from the medial geniculate nucleus and is located on parts of the superior temporal gyrus of the cerebral hemispheres (figure 7.10). The primary auditory cortex is organized according to tone, with low frequency sounds represented anteriorly and laterally, and high frequency sounds represented posteriorly and medially.

## The Olfactory Pathway

This is the only sensory input that does not project to the thalamus, but goes directly to the temporal cortex. Olfactory receptors are located in the lining of the posterior part of the nasal cavity. These cells project to the olfactory bulb, where their terminal axons end upon the dendrites of mitral cells in structures known as olfactory glomeruli. Mitral cells are the principal output cells from the olfactory bulb, and project via the olfactory tract to the olfactory regions of the forebrain. These olfactory areas include the primary olfactory cortex on the medial parts of the temporal lobe and the amygdala (a limbic system nucleus deep within the temporal lobe — see figures 7.11–7.12).

## The Brainstem and Input Processing

The brainstem plays a major role in the transmission and processing of sensory information or input, particularly with regard to the sensory modalities of touch, pain and hearing. Many of the ascending pathways to higher regions of the brain have either collateral, or quite independent connections to brainstem structures, and these connections are important for a variety of functions. These include: the modulation of pain perception via the actions of descending projections to the spinal cord; coordination of motor activation via proprioceptive input to the cerebellum; control of visceral reflex function via autonomic nuclei in the brainstem; most importantly for higher center function, modification of cortical function by ascending projections from the reticular activating system.

## The Thalamus as an Input Relay Station

From the discussion of the pathways for the major modalities given above, it will be clear that the thalamus plays a major role as a relay station for many ascending projections. The thalamic nuclei are many, but the role of this structure may be easily understood by considering functional groups.

Thalamic nuclei involved in the pathways for touch, pain, hearing and vision are considered as *specific* relay nuclei, projecting to discrete functional regions in the cortex concerned with each of those specific senses. Another group consisting of the pulvinar, anterior and dorsomedial nuclei are considered as *association* nuclei because they project to areas of the cortex with association function (parieto-temporo-occipital, limbic and prefrontal association cortices). A third group of thalamic nuclei are known as *non-specific* thalamic nuclei (e.g. intralaminar nuclei) because they project to widespread regions of the cortex.

## OUTPUT EFFECTOR SYSTEMS
## Somatic Motor Pathways

Voluntary motor commands are initiated within the motor areas of the frontal lobe. Large neurons located within the primary motor cortex on the precentral gyrus and anterior parts of the paracentral lobule send axons through the deep white matter of the cerebrum (internal capsule) to either the brainstem (corticobulbar fibers) or the spinal cord (corticospinal fibers). These fibers cross the midline during their course so that the left side of the cerebral cortex controls the right side of the face and body and vice versa. These direct cortical projections are concerned principally with fine control of independent movements, particularly in the hand. Other pathways concerned with motor control arise from the vestibular nuclei of the brainstem (vestibulospinal tracts) which are involved with maintaining balance and posture or the reticular nuclei of the pons and medulla (reticulospinal tracts) concerned with initiation of stereotyped movements (walking, swimming, etc) — see figure 7.10.

## The Basal Ganglia and Cerebellum

In addition to the above pathways controlling motor activity in the brainstem and spinal cord, there are several nuclei in the forebrain and midbrain concerned with motor function. These include the corpus striatum (caudate, putamen and globus pallidus) and associated nuclei which have reciprocal connections with them (substantia nigra in the midbrain, subthalamic nucleus in the diencephalon) — see figure 1.3 of The Big Picture and figure 7.6, 7.8, 7.9, 7.11.

The cerebellum also plays an important role in motor control (figure 7.5). One part (flocculonodular lobe) receives information about the position of the body in space from the vestibular system and uses that information to control the muscles concerned with posture of the body. Other parts receive information about joint position and muscle tension from the spinal cord (medial parts of cerebellar cortex), or are involved in feedback loops with the motor cortex (lateral parts of cerebellar cortex) and control synergy of motor function i.e. the smooth activation of appropriate muscle groups in the appropriate sequence.

## The Hypothalamus and Brainstem Reticular Formation

### Control of Endocrine and Autonomic Nervous System Function:

There are many aspects of body function which are not under direct voluntary control. The hypothalamus has a pivotal role in these processes via two groups of pathways (figure 7.11). The first of these involves the pituitary (or hypophysis), which is a small organ located beneath the diencephalon. Its anterior lobe is involved in the control of growth, metabolism and sexual function, while the posterior lobe is concerned with fluid balance, milk ejection from the breast and uterine contraction during birth. Both these parts exert their control on the appropriate organs by releasing hormones into the bloodstream. The hypothalamus controls the anterior lobe by releasing hormones into a delicate system of vessels connecting the two (the hypothalamo-hypophyseal portal system) and controls the posterior lobe by direct axonal connections (the hypothalamo-hypophyseal tract).

The hypothalamus also contains nuclei concerned with the control of eating and cardiovascular function. They exert their effects via connections with nuclei located in the reticular formation of the brainstem. The reticular formation nuclei are complex and contain regions concerned with the drive to breathe, the control of heart rate and blood pressure, the control of digestion, as well as some programmed movements such as walking, running and swimming. The sensory input to the reticular formation is derived from multiple modalities (e.g. pain, touch, hearing, balance and vision).

## Speech

Speech is an important component of behavior. We know, from studies correlating cerebral lesions with speech deficits, that this function depends upon two major cortical functional areas and a connecting pathway. These areas are usually found in the left cerebral hemisphere in both right-handed and left-handed individuals. Broca's area is located on the inferior frontal gyrus (figure 7.4) and is concerned with the expressive aspects of speech. Wernicke's area, concerned with the comprehension of speech, is located along the posterior parts of the superior temporal gyrus (planum temporale) and/or the adjacent gyri of the parietal lobe (supramarginal and angular). Functional studies indicate that there is considerable variation between individuals as to the precise location of Wernicke's area.

Wernicke's area receives input from the auditory cortex and projects to Broca's area via the arcuate fasciculus. Broca's area in turn has connections to the motor areas of the cortex. Thus, both these areas receive and make connections appropriate to the sensory-motor loops necessary for language function.

## ASSOCIATION FUNCTIONS OF THE HUMAN BRAIN

Much of the integrative functions of the human central nervous system take place within the cerebral cortex and any model should consider the anatomy of this part of the brain in detail. Several identified association areas have been described. These include the prefrontal, parieto-temporo-occipital and limbic association areas. The first two of these consist of discrete regions within the neocortex alone, while the latter consists of orbitofrontal cortical regions and associated thalamic, hypothalamic and limbic structures.

We shall be considering in detail the anatomy of the limbic system and cerebral cortex in subsequent sections, as well as the connections between and within the two hemispheres.

## STRUCTURAL ORGANIZATION OF THE LIMBIC SYSTEM
### Concept and Role of the Limbic System

The limbic system can be considered as one of the association systems in the cerebrum. It includes diverse areas of the brain which are interconnected by prominent fibre bundles, and which collectively appear to be involved in generating goals and processing emotions, memory and olfaction. This system also has a substantial influence on visceral functions through the autonomic nervous system.

### Basic Organization

Definitions of the limbic system differ, but the commonest schema includes the hippocampal formation, parahippocampal gyrus and cingulate gyrus of the cortex, as well as the amygdala (located deep within the temporal lobe), septal area (near the midline in the forebrain), hypothalamus (in particular two structures known as the mamillary bodies) and the anterior nucleus of the thalamus. These structures are interconnected by prominent fibre bundles (fornix, mamillothalamic tract, stria terminalis). According to one estimate, the fornix alone contains a fibre population of about 1.2 million axons. The limbic system is increasingly seen as an integrated circuit (Armstrong, 1991) — see figure 1.4 in The Big Picture and 7.12.

The hippocampal formation is a particularly important component of the limbic system, receiving input from widespread areas of the association cortex via the anterior end of the parahippocampal gyrus. It consists of the hippocampus, dentate gyrus, and part of the adjacent parahippocampal gyrus known as the subiculum.

### Sources of Input and Output Pathways

The limbic system in general receives its main input from the cortex and the olfactory system. It is consequently in a position to monitor activities in diverse areas of the cerebral cortex which are relevant to emotions and memory and to participate in emotional expression. The entorhinal cortex is the conduit through which mutimodal sensory information reaches the hippocampus. Output from the system is partly to the neocortex and partly to those areas of the brain concerned with the physical manifestations of behavior (hypothalamus, brainstem autonomic reticular nuclei, raphe nuclei and the spinal cord).

## STRUCTURAL ORGANIZATION OF THE CEREBRAL CORTEX

The human cerebral cortex has a surface area of about 2,430 cm$^2$ once all folds are smoothed out, and an average cortical volume of approximately 935 cm$^3$. This represents about 67% of total brain volume. The number of neurons is between 2.6 x $10^9$ and 16 x $10^9$ (Pakkenberg, 1966; Sholl, 1956; Zilles, 1990) and there are approximately $10^7$ neurons under each cm$^2$ of cortical surface, except in the primary visual cortex where packing density is 2.5 times greater (Nolte, 1993). Mean cortical grey matter thickness is approximately 0.28 cm in humans, but varies greatly according to region.

## Cortical Topography

We have already mentioned the several areas of cerebral cortex with sensory input or motor output functions. In addition there is widespread association cortex. We shall consider the functional topography of the cerebral cortex in detail here, lobe by lobe.

The frontal lobe contains several motor areas: the primary motor cortex and the premotor cortex (for initiation of motor commands), the frontal eye fields (which control the movements of both eyes in concert), and the supplementary motor cortex (which controls postural movements involving groups of muscles on both sides of the body). Other important areas are the prefrontal cortex (concerned with the control of socially appropriate behavior) and if the hemisphere is the dominant one (usually the left in both left and right handed people), the motor speech area (or Broca's area) which contains the motor programs for the generation of language. A small gustatory (taste) area may be located posterior to Broca's area.

The parietal lobe contains the primary somatosensory cortex anteriorly, and if the hemisphere in question is dominant, Wernicke's area, which is concerned with the comprehension of language. Most of the remainder of the parietal lobe is occupied by association cortex which is concerned with recognition of objects and faces and perception of three-dimensional space.

The temporal lobe contains the primary auditory cortex, located on the superior temporal gyrus and extending towards the midline along the surface of the transverse temporal gyri, and some parts of Wernicke's area in some individuals. On its lower aspect lies the primary olfactory cortex and the olfactory association cortex (for recognition of odors). Deep within the temporal lobe lies the limbic cortical hippocampal formation.

The occipital lobe contains the primary visual cortex and the visual association cortex. Within the latter areas lie specialized regions concerned with the perception of color, motion and form.

## Cortical Connections

Three basic types of fibers may be identified connecting the cerebral cortex with other parts of itself and other structures. These are: association fibers which interconnect cortical areas on the same side of the brain; commissural fibers which join cortical areas in opposite hemispheres; projection fibers which consist of fibers carrying information between the cerebral cortex and lower parts of the CNS.

Association fibers may be short, joining adjacent cortical regions, or long, joining distant cortical regions. Short association fibers (short arcuate fibers) run for distances of 2–5 cm between adjacent gyri. The long association fibers form several tracts including: the superior longitudinal fasciculus which runs between the frontal and occipital poles of the cerebrum (reflected as the large mauve arrow in the figures in The Big Picture); the arcuate fasciculus which arches up from the temporal lobe to join the superior longitudinal fasciculus; the inferior occipitofrontal fasciculus which joins the frontal and occipital poles of the cerebrum via the temporal lobe; the uncinate fasciculus which connects the temporal and frontal lobes. Another important association tract on the medial side of the cerebrum is the cingulum, which joins the cingulate cortex to the parahippocampal gyrus in the temporal lobe, and is an important part of the limbic system.

The largest commissural fibers joining the two hemispheres is the corpus callosum, which contains fibers joining the frontal, parietal and occipital lobes; the anterior commissure joins olfactory as well as other areas located in the temporal lobe. In old world primates, about 93% of commissural fibers cross in the corpus callosum, 5.3% cross in the anterior commissure while most of the remainder cross in the smaller hippocampal commissure associated with the limbic system (see below).

It is not possible to determine accurately the number of axons involved in these pathways as postmortem human tissue is rarely sufficiently well preserved for electron microscopy. However, rough estimates can be made based on the known fiber density in the human corpus callosum (200 million myelinated and unmyelinated axons in 430 mm$^2$, or about 0.47 million per mm$^2$ in cross-section; Aboitiz et al., 1992). The human anterior commissure has an approximate cross-sectional area of 10–14 mm$^2$ and 5–7 million axons. The uncinate fasciculus has a slightly larger cross-sectional area, suggesting a fiber count of the order of 10 million. The inferior occipitofrontal fasciculus in the human has a cross-sectional area of approximately 12–18 mm$^2$, yielding a value of about 6–9 million fibers. The superior longitudinal fasciculus is very large with a cross-sectional area of the order of 50–70 mm$^2$ and about 24–33 million axons, and at least half of these would appear from dissection to contribute to the arcuate fasciculus. Both the cingulum and superior occipitofrontal fasciculus are considerably smaller (approximate cross-sectional areas of 20 mm$^2$ or 9–10 million axons). These figures should be taken as very approximate only, but provide an example of further anatomical model parameters, in extending numerical brain simulations such as outlined in chapter 10a.

Projection fibers include the descending axons from the motor centers in the cortex (e.g. corticobulbar and corticospinal fibers), as well as ascending pathways to the primary sensory areas of the cortex. Numbers of these are very difficult to estimate as the fibers mingle in the internal capsule. Corticospinal fibers in humans are estimated to number approximately 1 million (Gray's Anatomy, 1989).

## Neuronal Organization of the Cortex

Most of the cerebral cortex has a 6-layered structure. These layers are numbered in order from the outside to the inside. Each of these layers has characteristic types of neurons within it and layers also differ in the types of axons that come from the underlying white matter to terminate in them (table 7.3).

Layer I is closest to the pial surface and contains very few neurons. It is also called the molecular layer because the large number of fibers present give it a punctate or molecular appearance in sections stained for nerve fibers.

Layer II is known as the external granular layer and contains many small neurons, including both stellate and pyramidal neurons.

Layer III is known as the external pyramidal layer and its constituent neurons are predominantly pyramidal, extending their apical dendrites into layers I and II. These pyramidal neurons send their axons into the white matter.

Layer IV is the internal granular layer and consists of closely arranged stellate cells,

**TABLE 7.3.**
Layers of the cortex.

many of which receive connections from axons arising from the neurons of the thalamus. Stellate cells in turn connect with the apical dendrites of neurons having their somata in layers V or VI, or contact other stellate cells.

Layer V is called the internal pyramidal layer and contains the largest pyramidal neurons along with some stellate cells. These large pyramidal neurons project to other parts of the brain including the thalamus, brainstem and spinal cord.

Layer VI is the multiform layer and contains neurons of many different shapes. Axons arising from some of the neurons in this layer contribute to the association, commissural and projection fibers.

Most incoming fibers (from the specific thalamic nuclei) are distributed to layer IV and the deeper part of layer V.

There are important regional differences in the relative thicknesses of the above layers. These six layers can be identified throughout most of the cerebral cortex in areas known as homotypical cortex. In some areas, known as heterotypical cortex, the six layered pattern is obscured by regional specializations. For example, in the primary visual cortex and in some parts of the auditory and somatosensory cortex, the stellate cells are so numerous as to overflow from layer IV into the adjacent layers. In these areas, known as granular cortex, layers II to V contain small neurons receiving fibers from the appropriate sensory parts of the thalamus. By contrast, in the motor and premotor cortex, layers II to V appear to contain large numbers of pyramidal neurons. These areas are known as agranular cortex.

Overall the pyramidal neurons are by far the commonest type, comprising about two thirds of all cortical neurons even in sensory areas. Most of the other neurons are stellate interneurons. In one study of part of the motor cortex (precentral gyrus; Cragg, 1967), each pyramidal neuron was estimated to contact about 600 other neurons, taking part in 60,000 synapses. In the visual cortex, where about one tenth of cortical neurons are concentrated, the dendrites of a single neuron may connect with 2,000 to 4,000 other neurons and an incoming projection fibre may contact up to 5,000 neurons (Gray's Anatomy, 1989).

The cortex is divided into columnar modules spanning the entire thickness of the grey matter. These cylindrical modules are believed to be based on cortico-cortical connections and are thought to be 0.2–0.3 mm in diameter. Human cortex contains about 2 million of these modules. About 70% of neurons in each cortical module send axons substantial distances from the locale of their soma (i.e. at least one cortical module width, or more than 0.3 mm). About half of these are directed to other cortical areas, and half project to subcortical regions (Szentagothai, 1986).

Several authors have subdivided the cerebral cortex into different areas on the basis of topographical differences in appearance in fiber and cell stains. The most famous and enduring of these schemes was produced by Brodmann and consists of 52 separate regions, many of which correspond to different functional areas of the cortex. Thus the primary visual cortex corresponds to Brodmann's area 17 and the primary somatosensory cortex to Brodmann's areas 1, 2 and 3, and the primary motor cortex to Brodmann's area 4. Figure 7.7 is a colour version of Brodmann's areas.

## Neurotransmitters in the Cortex

Large pyramidal neurons of the cortex have been shown to contain the amino acid neurotransmitter glutamate, which appears to be the main transmitter for brief, direct, excitatory synaptic connections in the CNS as a whole. Gamma aminobutyric acid (GABA) is a common inhibitory neurotransmitter in cortical interneurons.

### The brain's 'chemoarchitecture'

There are important connections arising from the brainstem and basal forebrain which influence diverse areas of the cerebral cortex and which have been implicated in sensory attention, learning and mood. These connections are, of course, likely to be of major importance to our understanding of emotional behavior and mental illness. We may consider these diverse connections on the basis of the neurotransmitters used (see The Big Picture, chapter 6 and figure 7.13).

Pathways containing the neurotransmitter acetylcholine (cholinergic pathways) arise from nuclei in the base of the forebrain (between the hypothalamus and the orbital cortex), in the basal nucleus of Meynert and part of the septal nuclei, and are distributed to the cerebral cortex, hippocampal formation and amygdala. These cholinergic neurons, together with the brainstem reticular formation, may play a role in the sleep-wakefulness cycle and memory. Quantitative data concerning the total number of cholinergic neurons in the basal forebrain has been estimated to be 200,000 to 415,000.

Serotonergic pathways arise from the raphe nuclei in the midline of the brainstem. Rostral raphe nuclei (e.g. dorsal raphe nucleus) project to virtually every part of the forebrain (including thalamus, hypothalamus, striatum, hippocampus, amygdala and all neocortical areas). The cortical innervation is most dense in sensory and limbic areas. It has been estimated that the human dorsal raphe nucleus contains 235,000 neurons of which most are serotonergic. More caudal raphe nuclei (e.g. nucleus raphe magnus) project to the spinal cord and cerebellum. There are approximately 78,000 serotonergic neurons in the human medulla. Ascending projections may modulate the general activity of the CNS (e.g. in determining the level of arousal and mood), while descending projections may modulate pain.

Pathways using the neurotransmitter dopamine are shown in figure 7.13. Neurons of origin lie in the midbrain. These pathways are concerned with motor function (e.g. the pathway from the substantia nigra to the striatum), but also distribute to the cortex (mainly from the ventral tegmental area to motor and limbic cortical areas). The dopaminergic pars compacta subregion of the human substantia nigra is estimated to contain 200,000 to 270,000 neurons (van Domburg and ten Donkelaar, 1991), while the ventral tegmental area has 50,000 to 80,000 neurons (van Domburg and ten Donkelaar, 1991). Projections to limbic areas suggest that these pathways may be involved in motivation, reward and cognition as well as initiation of movements.

A small but important nucleus in the brainstem known as the locus coeruleus gives rise to axons containing the neurotransmitter norepinephrine, which are distributed to many areas of the cerebrum, including thalamus, hypothalamus and cerebral cortex. The human locus coeruleus has approximately 54,000 noradrenergic neurons, with a further 6,260 noradrenergic neurons located in the adjacent subcoeruleus region. The noradrenergic cortical projections may be important in arousal, vigilance, attention, and 'fight and flight' processing.

Neurons containing the amino acid excitatory neurotransmitter glutamate include interneurons in many parts of the CNS and large cortical pyramidal neurons projecting to the striatum, thalamus and lower motor neurons. Inhibitory GABA is found within many interneurons, Purkinje cells of the cerebellum, the striatum, globus pallidus and some thalamic nuclei.

## NEUROANATOMY AND BRAIN MODELING

For the above discussion we have neatly divided the central nervous system into apparently discrete systems. This may have left the reader with the impression that the functional activity of the brain can also be neatly parcellated. Day to day experience would indicate that this is clearly an oversimplification, since the minute by minute behavior of humans depends upon the integrated activity of almost all the above systems. We might expect, then, that there would be abundant interconnections between the individual systems. These would most logically occur at the level of the cerebral cortex via the association and commissural connections discussed above.

Formulation of useful models of human brain function calls for detailed anatomical and chemical information. It should be clear to the reader from the above account that a great deal of anatomical information is known, but the state of knowledge of human neuroanatomy allows only vague answers to such questions as the number of fibers in major tracts or the precise patterns of connections between them. Part of the problem lies in the difficulty of obtaining human postmortem material in a satisfactory state to permit electron microscopy. Without such an analysis it is impossible to give accurate estimates of the populations of unmyelinated axons within most fibre tracts and hence the total number of axons in the major pathways. This problem is further compounded by the lack of precise boundaries to many of the major association bundles in the human brain. New generation (for example carbocyanine) dyes may provide the answer, but these postmortem tracers are currently difficult to use in mature tissue and are not particularly suited to quantitative studies.

A multidisciplinary approach to brain science means that the detail about brain anatomy needed by brain modelers must become better known to specialist anatomists. Such information exchange will lead to an increasing convergence of anatomical and functional perspectives in brain science.

## REFERENCES AND SUGGESTED FURTHER READING

Aboitiz, F., Scheibel, R. S., Fisher, E. and Zaidel. (1992) Fiber composition of the human corpus callosum. *Brain Research,* **598,** 143–153.

Armstrong, E. (1991) The limbic system and culture. *Human Nature,* **2**(2),117–136.

Cragg, B. G., (1967) The density of synapses and neurons in the motor and visual areas of the cerebral cortex. *J. Anat.* **101,** 639–654.

Cragg, B. G., (1976) Ultrastructural features of human cerebral cortex. *J. Anat.* **121,** 331–362.

Van Domburg, P. H. M. F. and H. J. ten Donkelaar (1991) The human substantia nigra and ventral tegmental area. A neuroanatomical study with notes on aging and aging diseases. *Advances in Anatomy, Embryology and Cell Biology.* Vol 121. Springer-Verlag. Berlin.

*Gray's Anatomy.* (1989) Edited by P. L. Williams, R. Warwick, M. Dyson and L. H. Bannister. 37th Edition. Edinburgh. Churchill Livingstone.

Nolte, J., (1993) *The human brain. An introduction to its functional anatomy.* 3rd Edition. St. Louis, Mosby.

Pakkenberg, H. (1966) The number of nerve cells in the cerebral cortex of man. *Journal Comp. Neurol.* **128,** 17–20.

Sholl, D. A. (1956) The measurable parameters of the cerebral cortex and their significance on its organization. In *Progress in Neurobiology.* Edited by J. A. Kappers (pp 324–333). Elsevier. Amsterdam.

Szentagothai, J. (1986) The architecture of neural centers and understanding neural organization. In *Advances in Physiological Research.* Edited by H. McLennan, J. R. Ledsome, C. H. S. McIntosh and D. R. Jones. New York, Plenum.

Zilles, K. (1990) Cortex. Chapter **22** (pp 757–802) In *The Human Nervous System.* Edited by G. Paxinos. Academic, San Diego.

# Chapter 8

# SENSORY-MOTOR MODELS OF THE BRAIN

Alan Freeman

Department of Biomedical Sciences
The University of Sydney

In even the simplest view of the brain, its sensory and motor capabilities figure prominently. Ask someone what the brain does, and the answer might well be that the brain lets you hear, see, feel, taste, and smell, and allows you to act on such sensory information. Phylogenetically more recent portions of the brain were built around a core of sensory and motor pathways.

This chapter explores the nature of sensory-motor systems in three steps. The simplest model is that of a sensory-motor reflex loop, that produces stereotyped actions in response to sensory environmental stimuli. Next, the sensory pathway is treated as a series of filters that pick out those aspects of the incoming information that are of most relevance. The motor pathway is then described as a series of transformations that translate an overall plan for action into contractions of individual muscles. Finally, a dynamical perspective of sensory-motor processing is presented.

## BEHAVIOR AS A REFLEX

A sensory stimulus commonly elicits a standard reaction. A loud or unusual noise produces a turn of the head and eyes. A painful stimulus to a finger produces a rapid withdrawal of the arm. These are examples of sensory-motor reflexes, in which sensory nerve cells convey information to motor spinal cord and brain networks, which innervate muscles in a more or less automatic fashion. These reflex arcs provide us with a useful starting point, since the more sophisticated functions of both the sensory and motor systems (and aspects of cognition) are built upon these simple and phylogenetically ancient reflex pathways.

Reflexes are ubiquitous in the nervous system, regulating everything from blood pressure to contraction of the urinary bladder. We will examine just one set of reflexes, those that maintain body posture. These provide a useful example because they range hierarchically from the simplest to some of the most complex reflexes in the nervous system.

## Spinal Reflexes

Probably the simplest reflex of all is the stretch reflex: rapidly stretching a muscle elicits an equally rapid muscle contraction. A particularly reliable example of the stretch reflex is the knee jerk. The subject sits with the lower leg hanging free. Tapping just below the knee results in a sudden forward movement of the lower leg. What is happening here? The muscle involved is the quadriceps (an upper leg muscle with a tendon that inserts adjacent to the knee-cap), and tapping just below the knee pushes on the tendon and stretches the muscle. Embedded in the muscle are neurons responsive to stretch, which increase their action potential rate in response to the stretch, and convey this signal along their axons to the spinal cord motor neurons, whose increased rate of action potentials result in muscular contraction.

The stretch reflex helps to maintain a fixed body posture. Consider, for example, standing for a long period. The leg muscles keeping the knee straight will tend to tire. Any bending of the knee will tend to stretch the knee extensor muscles, leading to reflex contraction and straightening of the knee. More generally, any change in body posture will be resisted by the mechanisms of the stretch reflex. Another important example of a spinal reflex is the withdrawal reflex in which a potentially painful stimulus to a limb (skin receptors) results in connection to motor neurons that rapidly flex the limb and withdraw from the stimulus.

The spinal reflexes can be modulated by a number of influences from elsewhere in the nervous system. If a person is to walk, for instance, it is obviously inappropriate for the stretch reflex to keep the knee straight. The influence of higher centers on the stretch reflex can be illustrated by Jendrassik's maneuver. The subject hooks one hand onto the other in front of the chest and tries to pull the hands apart. This usually increases the size of the kick produced by the knee jerk. Sensory neurons are clearly not providing the only synapses onto the motor neuron subserving the stretch reflex. There are also other synapses provided by neurons in higher centers, whose activation increase the likelihood of motor neuron activity during the maneuver. This observation leads us to proceed to the next level of postural reflex, that residing in the brainstem (medulla, pons and midbrain — see the previous chapter, The Brain's Anatomy).

## Brainstem Reflexes

Early this century, Charles Sherrington described a series of experiments showing the major role that the brainstem plays in controlling body posture (Sherrington, 1961). His basic method was to surgically divide the brainstem in cats, and to observe the resulting posture. The first result was that separation of the brainstem from the spinal cord resulted in an immediate loss of all spinal reflexes. Evidently motor centers higher than the spinal cord provide a net excitatory influence to motor neurons, and this facilitation is necessary for normal spinal reflexes to take place. The same result is seen when the upper spinal cord is suddenly transected (for example, through trauma) in humans: there is a loss of all contraction in muscles driven by motor neurons below the level of the lesion.

Sherrington next showed that surgical separation of the midbrain from the pons (decerebration) produces a very different result from spinal transection: all four limbs (in cats) were stiffly extended. Decerebrate rigidity has also been seen in humans. In this case both arms and legs are extended, and the back and neck are arched. It is apparent from these observations that motor areas in the medulla and pons of the normal subject provide a net excitation to extensor motor neurons, and that the resulting extension is sufficient to support the subject against gravity. In general, then, the lower brainstem is providing an antigravity stance.

Decerebrate posture is inflexible in that it cannot adapt the subject to its environment. A decerebrate cat placed on a sloping surface falls over and cannot right itself. These postural deficiencies are rectified when the brainstem lesion is placed above the midbrain rather than below it. In this case, termed a decorticate lesion, the spinal cord and brainstem are intact but are not connected to higher motor centers. A decorticate animal still supports itself against gravity, but adjusts the extension of its limbs to an uneven surface and can right itself if it falls. Sensory input is clearly being used by the midbrain to adjust posture to suit the environment.

In summary, the lower brainstem provides a stiff antigravity stance and the upper brainstem moderates this into an flexible antigravity stance.

## Cortical Reflexes

The highest level reflex belongs to somatosensory and motor cortex. Whereas the output of the brainstem is directed to motor neurons driving the muscles of the limbs, body, and neck, cortical output is directed primarily to the motor neurons driving distal muscles (in particular the muscles of the hand). This reflex is therefore concerned with fine, skilled, and rapid movement of the distal musculature. For example, in the process of writing, the brainstem is able to set the background posture of the limbs, body, and neck, against which the somatosensory-motor cortex can drive the fine movements of the fingers.

The primary somatosensory and motor cortices have a very close relationship, well-suited to the needs of a sensory-motor reflex loop (see figures 1.2 and 1.3 in The Big Picture). They share a common border (the central sulcus), and cells on the sensory side are richly connected to those on the motor side. Neurons in primary motor cortex can be activated by stimuli near the muscles driven by those neurons. For instance, they receive input from receptors in the muscles they drive, and from receptors in areas of skin adjacent to the driven muscles. Thus there is a transcortical reflex loop comprising a sensory pathway, motor cortical cell, and motor neuron, that is analogous to the spinal reflex loop. Matthews (1991) described some of the evidence that the transcortical loop modulates lower-level reflexes to fit current postural requirements. For instance, when an object is being carried in the hand with the palm turned upwards, the biceps stretch reflex is active and the antagonist of the biceps muscle, the triceps, is relaxed. But when the palm is turned downwards, biceps and triceps need to be simultaneously active and both biceps and triceps stretch reflexes can be elicited. The loop running through the cortex is apparently modulating the spinal loop.

## Behavior is More Than A Reflex

We have thus far viewed sensory input as a prompt for motor output, and conversely, motor activity as dependent on a sensory trigger. This is a view taken to its extreme by behavioral psychologists such as B. F. Skinner, who argued that many aspects of behavior could be seen as the outcome of a huge sensory-motor reflex (Skinner,

1938). This model is clearly not sufficient to account for much of what the brain does. We can have sensory experience in the absence of any movement, and movement without sensory experience. A number of models have been outlined in The Big Picture (for example Sokolov and Gray) that focus on the decision making and evaluative dimensions underlying sensory-motor processes. It is therefore appropriate to consider next the nature of the sensory and motor systems as separate and potentially independent entities.

## SENSORY SYSTEM

The sensory systems perform three major functions: filtering information from the environment, combination of the filtered streams of information, and comparison of the result with previously stored information. To illustrate the filtering function, consider the example of electromagnetic energy. Such energy is carried by waves whose wavelengths vary from about 10–15 m to $10^8$ m. Our bodies are continually bombarded by radiation from much of this spectrum, whether it be X-rays and ultraviolet light from the sun, or the warmth of a fellow human. Yet we are sensitive to only a tiny fraction of this spectrum, from about $4 \times 10^7$ m to $1 \times 10^4$ m. The lower wavelengths in this range are sensed with the eyes, and the upper wavelengths with the thermoreceptors of the skin. Information carried by the vast bulk of the electromagnetic spectrum is outside this range and is discarded, presumably because it does not significantly affect our chances of survival.

Secondly, there is the process of combination. The input to this process comes from small portions of the sensory surface which each analyze part of the stimulus. The combinatorial process puts together these fragmentary descriptions to form a more global description of the stimulus, such as its speed, direction, or shape. An example is the auditory system, where input from the two ears is combined at several stages of processing. One of the major outcomes of this binaural interaction is the ability to localize the source of a sound. Thirdly, at the highest levels of the sensory systems, incoming information is compared with stored images, in the process of object recognition. Input from all the sensory systems comes together at this level to form a polysensory image. Thus, the whiff of a sand dune can bring back memories of the sights and sounds of a previous seaside holiday.

### Serial Processing

The filtering, combination and comparison processes can be broken down into a series of transformations, as shown in table 8.1. These stages are described in sequence, using examples from the major sensory systems. The most important anatomical sites for these stages are also shown in the table for the auditory, somatosensory and visual systems.

### Stimulus Conditioning

At the entrance to the neural pathway of each sensory system is an array of receptors, cells that turn environmental energy into neural energy. Between the receptors and the outside world is a layer of non-neural tissue that conditions the stimulus into a form suitable for the receptors. This tissue maximizes the transmission of certain

TABLE 8.1.

Filtering, combination and comparison in sensory systems.

| Modality | Stimulus conditioning | Transduction | Adaptation | Lateral antagonism | Relay modulation | Feature Building | Selective attention, object recognition |
|---|---|---|---|---|---|---|---|
| Vision: | Optics | Photo-receptors | Receptor, outer plexiform layer | Retina | Lateral geniculate nucleus (of thalamus) | Primary visual cortex | Visual cortex |
| Hearing: | Outer, middle inner ear | Inner hair cells | Hair cells | Cochlea | All subcortical nuclei | Subcortical nuclei | Auditory cortex |
| Touch: | Epidermis, dermis | Low threshold mechano-receptors | Skin, receptor capsule | Dorsal column nuclei | Ventral posterior nucleus (of thalamus) | Primary somato-sensory cortex | Somato-sensory cortex |

TABLE 8.2.

Transformations in every movement.

| Type of movement | Locating the target | Planning the action | Fine movement control | Postural control | Low-level coordination | Driving individual muscles | Muscle |
|---|---|---|---|---|---|---|---|
| Fine movement: | Posterior parietal cortex | Premotor cortex | Primary motor cortex | | Cranial nerve nuclei, spinal cord | Lateral motor neurons | Distal muscles |
| Postural adjustment: | Posterior parietal cortex | Premotor cortex | Primary motor cortex | Brainstem postural centers | Cranial nerve nuclei, spinal cord | Medial motor neurons | Proximal muscles |
| Touch: | Posterior parietal cortex | Frontal eye fields | Brainstem gaze centers | Integrator in cerebellum, brainstem | Cranial nerve nuclei | Ocular motor neurons | Extraocular muscles |

forms of energy to receptors, and other forms of energy are transmitted poorly or not at all. In the ear, the small bones of the middle ear efficiently transmit longitudinal vibrations of air (sound) to the watery fluid bathing the auditory receptors by matching the resistance of the eardrum to that of the fluid. In the absence of the matching process most of the sound would be reflected back out of the ear without activating the receptors. The ocular lens transmits those wavelengths of light to which the receptors are most responsive, but absorbs ultraviolet light which could damage the delicate structures of the retina.

The non-neural tissues between the external world and the receptors have a second major function: dividing the stimulus up so that each receptor sees only a tiny portion of it. Receptors are laid out in flat sheets. The conditioning process spreads the stimulus across the sheet so that each receptor reports only on that bit of the stimulus falling on its portion of the sheet. In the ear, the receptor sheet runs along the length of the cochlea. High frequencies of sound activate receptors mainly at one end of the cochlea, while low frequencies activate receptors at the other end. In the eye, the optical apparatus has the job of concentrating light from each point in visual space onto a corresponding location on the retina. For each retinal point, the imaging process maximizes the transmission of light from one point in space and minimizes the transmission of light from elsewhere. The receptors, which are arrayed across the retina, therefore each report on one small fragment of visual space.

## Transduction

Transduction is the process responsible for transforming the energy present in the environment into the electrochemical energy used by nerves. The hair cells in the inner ear, the touch receptors in skin, and the stretch receptors in muscle, for instance, all turn mechanical distortions in tissues that surround them into changes in the voltage across their cell walls. It is the transduction process that largely defines the spectral filtering in a sensory system. Hair cells in the cochlea respond little to sounds with frequencies outside the range 20–20,000 Hz, cutaneous mechanoreceptors respond poorly to skin vibration above about 500 Hz, and photoreceptors are sensitive only within the wavelength range 400–700 nm.

## Adaptation

Receptors typically respond transiently to a constant stimulus. This is the process of adaptation, which means that a receptor gives a relatively vigorous response at the start of a stimulus, followed by a declining response even though the stimulus remains unchanged. With the exception of the receptors responsible for pain, all receptors adapt. There is also adaptation in the nerve circuits beyond the receptors, but for the sake of brevity we will concentrate on the receptoral variety.

The process of adaptation confers on a sensory system two great advantages. First, an unchanging stimulus will rarely convey any information of survival value. It is better that the response to an unchanging stimulus be discarded, since if allowed to remain it may hinder the response to a changing, and therefore more relevant stimulus. Secondly, adaptation markedly expands the dynamic range of a sensory system. Consider the case of a visual scene viewed at different levels of ambient illumination. In the absence of adaptation, the highest level of illumination yielding useful vision would be only about a hundred times greater than the least level, due to the limited rate at which nerve cells can produce action potentials. In fact, a photoreceptor, and the retinal cells to which it connects, has a dynamic range closer to a million. This comes about because the process of visual adaptation desensitizes retinal cells, in particular the receptors. When illumination is increased by a certain factor, the receptors increase their responses by a lesser factor since they have lost sensitivity, and they therefore reach the limit of their ranges at a higher illumination.

## Lateral Antagonism

Each sensory cell has an area on the receptor sheet in which stimulation will change its output signal. This area is called the receptive field. Many cells in the somatosensory and visual systems respond well when focal and edge stimuli are directed at their receptive fields, but respond poorly when their fields are uniformly stimulated. This property, lateral antagonism, arises because the signals from receptors in the periphery of the receptive field are wired up so as to antagonize (that is, reduce) the signals from the receptive field center. It is a ubiquitous property of sensory systems and has been found in species ranging from horseshoe crabs to humans. Lateral antagonism presumably serves the requirements of object recognition circuitry higher up in the sensory system. An object is much easier to recognize from its corners, edges, and contours, than from its uniform surfaces. The sensory systems therefore discard information about spatially uniform stimuli in favor of responses to non-uniform stimuli. There is therefore an analogy between adaptation and lateral antagonism: the former reduces the response to stimuli constant in time, the latter to stimuli constant in space.

## Relay Modulation

In each of the major sensory systems there is at least one relay center between the receptors and the cerebral cortex. In the somatosensory system for instance, signals from the low-threshold mechanoreceptors pass through synaptic stations in the brainstem and thalamus before reaching the cortex. The connections made in these relays do not mix the signals coming from different receptor types, and they preserve the precision with which a stimulus can be located on the receptor sheet. They are susceptible, however, to modulation from signals feeding back from locations higher up the pathway, and to influences from outside the pathway. In our example of the somatosensory system, the somatosensory cortex feeds back to the synaptic stations in both brainstem and thalamus.

What are these modulatory signals doing? This question is largely unanswered, but two suggestions can be made in the case of the visual system. The relay nucleus lying between the retina and visual cortex is the dorsal lateral geniculate nucleus of the thalamus (L in figure 1.2 of The Big Picture). There are clearly modulatory influences occurring here, since only 10–20 % of the synaptic connections onto geniculate relay cells come from the retina. One source for the remaining synapses is the midbrain. It has been shown that the midbrain is responsible for facilitating relay cells during wakefulness and rapid-eye-movement sleep, and for reducing relay cell activity during synchronized slow-wave sleep (see Singer, 1977). Another source of synapses is the visual cortex, since just as many axons run from the visual cortex to the geniculate as run from the geniculate to the cortex. There is an indication from some recent work that the geniculate and cortex may be working together to shape cortical receptive fields (Sillito *et al.*, 1994). As a working hypothesis, therefore, relay modulation may be acting to reduce signal transmission through the relay, leading to periods of lowered awareness of sensory signals, and thereby helping to determine receptive field properties in higher centers.

## Feature Building

At this stage the discarding of information from the environment is more or less complete, the input has been sufficiently filtered, and a synthetic process commences. The receptive fields of cells providing the input for this stage are small: they collect information from just a few receptors, and therefore a very limited area of the receptor sheet. In the feature building stage, the inputs are combined so that receptive fields become bigger and information about location on the sensory sheet is diminished. Instead, the sensory systems build up signals representing relatively complex features of external objects such as their shape, and their distance and direction from the observer. The nervous system has evolved a number of beautiful strategies for this feature building process (Barlow, 1986), several of which will be described.

The input to the primary visual cortex is punctate, in that it describes a contrast in light levels at a specific place in the visual field. The primary visual cortex combines several neighboring inputs to produce a signal indicating the orientation of a contour or edge at that location in the visual field (see figure 8.1). Thus a vertical edge will excite one group of cortical cells, an edge inclined at 20° to the vertical will excite a different group of cells, and so on. Further, there is cortical circuitry to indicate the direction of motion of an edge. An edge passing a point in the visual field will excite a unique group of cells depending on the direction it is heading. To provide this signal, the cortex must be combining its inputs from at least two points along any potential path across the visual field. The primary visual cortex also synthesizes a signal corresponding to the distance of an object from the observer. Inputs to the cortex are essentially monocular: input from the two eyes differs because of the lateral separation between the eyes. These inputs are combined so that different cortical cells respond when a stimulus is closer than, at the same distance as, or beyond the fixated object.

As with the visual system, inputs to the somatosensory cortex are punctate. These inputs indicate, for instance, the presence of mechanical distortion on a small area of one finger. Within the somatosensory cortex, however, are cells that respond to mechanical stimulation over broader areas such as a whole hand. These cells do not respond well to punctate stimuli but instead to three-dimensional objects placed on or moved across the hand. The auditory system is unusual in that the combination of inputs occurs well before the sensory cortex. Inputs from the two ears are combined very soon after they enter the central nervous system, at the level of the brainstem. The intensity and timing of signals from the two ears are compared to determine the direction from the observer of a sound source.

## Selective Attention and Object Recognition

The primary sensory cortex sends part of its output to the higher sensory cortex. A large portion of the cerebral cortex falls into this category: in the case of the visual system, for instance, about half of the cerebral cortex is visual in function, and most

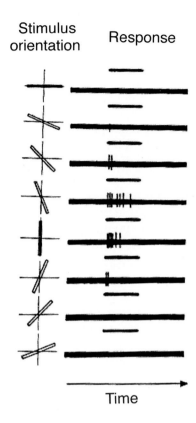

**Figure 8.1.** Orientation: Sensory cells at higher levels of the sensory pathways respond to increasingly complex aspects of the stimulus. While a receptor signals a stimulus at its location on the receptor sheet, higher order neurons signal stimulus features extending across many receptors. The illustration shows the discharge (each upstroke represents an action potential) of a cell in cat primary visual cortex when light bars of various orientations are presented to the receptive field. The cell responds best when the bar is inclined anticlockwise from vertical (Adapted from Hubel and Wiesel, 1959).

of that lies beyond the primary visual area. Despite this relatively large area of cortical surface, little is known about what it does. Three roles can be mentioned: further feature building, selective attention and object recognition.

An example of the feature building is motion discrimination. As described above, there are cells in the primary visual cortex that respond to stimulus motion. They respond to the component of motion in a direction perpendicular to their preferred stimulus orientation, and therefore cannot discriminate between stimuli having different directions of motion but the same velocity component in the preferred direction. Motion-sensitive cells in a higher area of cortex (MT), on the other hand, can make this discrimination and therefore represent an extra stage in the processing of image motion.

Two broad cortical pathways for visual information have been described (Mishkin, et al, 1983). The first, the dorsal pathway, passes from the occipital to the parietal lobe, and includes the MT area. It is concerned particularly with location and movement in space and plays a major role in stimulus finding and tracking. The second pathway, the ventral pathway, leads from the occipital to the temporal lobe and is more concerned with object shape and color, that is, with describing the form of an object. These two routes have been nicknamed the *where* and *what* pathways, respectively.

How does one attend to one object in the visual field at the expense of all others? Visual attention is not simply a matter of directing the eyes to the point of interest. As you direct your eyes towards these words, for instance, you can direct your attention elsewhere, to the corner of the book. Such selective attention presumably shines some sort of neural spotlight on the relevant signals in the dorsal pathway (up and to the left) and the ventral pathway (a corner shape), and provides some sort of link between these descriptions of the object of attention. There are currently more ideas than solid data on this topic. And how does one recognize an object? How does one know that the L-shaped thing up and to the left is a book corner? The incoming information is presumably compared with a stored image. This sort of function has conventionally been consigned to the place where the auditory, somatosensory, and visual cortices meet, the association cortex of the parietal lobe. Again, little is known about how this process operates.

## Parallel Processing

Table 8.1 may give the mistaken impression that sensory processing can be described as predominantly serial. This is far from the truth. The sensory systems operate in a massively parallel manner. We will look at three forms of parallel processing in the sensory systems: topographic mapping, spectral analysis, and a form peculiar to the visual system, the parvocellular/magnocellular dichotomy.

## Topographic Mapping

Each receptor, or small group of neighboring receptors, gives rise to a sensory channel that remains relatively independent of nearby channels for the journey from the sensory periphery to the primary sensory cortex. Such a channel reports on stimuli if and only if they are delivered to its own small area of the receptor sheet. Further, these channels are topographically organized. This means that cells that are neighbors in a sensory nucleus are connected to neighboring receptors. In turn, this means that the receptor sheet is mapped onto each sensory nucleus. In the auditory system, there is a channel from each point on the receptor sheet in the cochlea to primary auditory cortex. Consequently the mapping of the receptor sheet onto the cortex is said to be tonotopically organized. Similarly, the cortical maps of the skin and retina are arranged somatotopically and retinotopically, respectively.

As previously discussed, there is extensive convergence after primary sensory cortex so that topographic mapping becomes much weaker after this point. Conversely, the existence of topographic mapping in primary cortex presumably simplifies the circuitry in higher order cortex since the convergent connections can be made between neighboring inputs. If the brain is to recognize the shape of an object placed in the hand, for instance, input needs to be combined from channels serving cutaneous mechanoreceptors over much of the hand and from other channels originating in muscles serving the same area. The somatosensory pathways are organized so that all these channels are adjacent in primary somatosensory cortex.

## Spectral Analysis

Alongside the parallel pathways from receptors of the same type there are pathways from related receptors. The three types of cone photoreceptor in the retina provide the most familiar example of this parallel organization. At each point in the retina the three cone types spectrally analyze incoming light into low, medium and high wavelength components. There is some reorganization of these three channels further along the visual pathway, but the point remains that the cortex is provided with a three-dimensional spectral analysis of the stimulus at each retinal point: the result is the perception of color. There is a similar spectral analysis performed by the low-threshold mechanoreceptors of the skin. Slowly-adapting receptors respond best to low frequencies of skin vibration, rapidly adapting receptors to medium frequencies, and Pacinian corpuscles to high frequencies. This analysis, which is performed at each point on the skin, provides a basis for sensing the texture of surfaces.

## Magnocellular and Parvocellular Pathways

There is another type of parallelism that falls into neither of the above categories — part of the *where* and *what* pathways mentioned previously. It is peculiar to the visual system and will therefore be only briefly described. The output cells of the retina fall into two major groups, those that project to the magnocellular layers of the lateral geniculate nucleus and layer 4C of the cortex, and those that project to the parvocellular and 4C layers. These two pathways, however, are driven by the same set of photoreceptors. There is a clear functional distinction between the two pathways. The magnocellular pathway is superior in its ability to carry rapidly varying signals, plays a major role in the motion-detection mechanisms of the cortex, and contributes to the dorsal pathway of higher-order visual cortex (Merigan and Maunsell, 1993). The parvocellular pathway is specialized for signaling fine spatial details of the stimulus, carries almost all of the color information from the retina, and contributes to the ventral pathway.

## MOTOR SYSTEM

The motor system provides us with the means for making our mark on the world. In producing movement, the motor systems are faced with two conflicting requirements. On one hand, movement takes place against varying and often unpredictable loads and obstacles. To complete an action, the motor control systems must therefore use sensory feedback to check on the progress and accuracy of the action. On the other hand, movements often have to be made in too short a time to allow for feedback. In catching a ball, for instance, there is no time to observe the gap between ball and hand. Instead the brain has to predict the future path of the ball and extend the hand to intersect that path. This is an example of a feedforward control system. The motor systems use both feedforward and feedback control, which will be discussed in turn.

## Feedforward Control

The motor systems perform a remarkable analysis each time a movement is made. At the input end there is the command to move the body to achieve some end. In the ball-catching example, the catcher has to simultaneously turn head and shoulders to the approaching ball, run towards its destination, outstretch an arm and hand, and then wrap the fingers around the ball. At the output end the motor system has to provide a precise time course of contraction and relaxation over several seconds for each of hundreds of muscles. This analysis (in various networks) occurs through a series of transformations, a simplified representation for which is shown in table 8.2. The motor system has to break down a command involving the whole body into commands involving ever-smaller groups of muscle until the drive is specified for individual muscles. Each of the transformations will be examined in sequence.

## Locating the Target

Posterior parietal cortex provides a high-level interface between the sensory and motor systems. The sensory systems locate a target of interest (a ball, a glass of water, a pen) and provide information on its shape, location, and (if it is moving) direction and speed. This information is collected by the posterior parietal cortex in which, as previously described, auditory, somatosensory, visual, and other sensory signals come together to form the dorsal, or *where*, pathway. The posterior parietal cortex sends its signal forward to the premotor area of the frontal cortex to fashion the motor command. The responses of some posterior parietal cells depend on the attentive state. Some respond, for example, during a movement towards an object of interest, but do not respond during the same movement when the object is absent. Such cells could clearly play a role in the initiation of movement.

## Planning the Action

In front of the primary motor cortex is the premotor cortex (more precisely, the lateral portion of the strip is called premotor cortex and the medial portion is the supplementary motor area) — see figure 1.3 in The Big Picture. The premotor cortex plays a major role in the planning of motor activity (evidence comes from lesion studies, electrical stimulation and brain imaging studies in which subjects were asked to mentally rehearse the task without executing it — blood flow increased only in the premotor area).

Table 8.2 suggests that the output of the premotor cortex goes only to the motor cortex. This is an oversimplification. There is an information superhighway, the corticospinal tract, extending from the sensory-motor cortex to the brainstem and spinal cord. This tract conveys approximately equal number of axons from the premotor and motor cortices to lower motor areas. Thus the premotor cortex is directly connected not only to the primary motor cortex but also to motor areas such as the brainstem. As we have already seen, the brainstem plays a major role in adjusting body posture. The connection of the premotor cortex to the brainstem therefore allows the direct adjustment of posture by the premotor area.

## Fine Movement Control

The corticospinal tract evolved in two stages, a medial component followed by a lateral one. The medial system descends the spinal cord close to its midline and ends on or near motor neurons innervating proximal muscles (the muscles of the neck, body, and limbs). These are the muscles of the body's axis, and therefore of posture. The lateral system descends in the lateral spinal cord and targets motor neurons driving muscles in the hands and fingers (distal muscles). The lateral system therefore controls the muscles producing fine movement. The premotor area can drive both the medial and lateral systems, but while it can act on the medial system directly, it can only drive the lateral system via the primary motor cortex. In this sense, the primary motor cortex is the origin of the lateral corticospinal tract and is uniquely responsible for fine movements such as those of the fingers.

What is the relationship between the discharge of single corticospinal neurons and the contraction of individual muscles? Edward Evarts showed that a neuron's discharge correlates with the amount of force produced in the muscles it drives: when the muscle contracts against a load the neuron discharges rapidly, but when the same movement is made in the absence of a load there is little or no discharge (Evarts, 1968). It also requires a population of corticospinal neurons acting in concert to specify a unique direction of movement (Georgopoulos, 1986).

## Postural Control

Progressing downwards through the hierarchy of the motor control systems, we find that there is a steady shift from voluntary to relatively automatic control. Thus the brainstem postural centers are governed by a mixture of feedforward signals from the cortex, and feedback signals from a variety of receptors. The issue of interest here is the feedforward signal, which typically adjusts the position of the body's axial structures to accommodate movements of the limbs and hands. When the arms are used to pull a load towards the body, for instance, leg muscles contract before the arm muscles in order to stabilize the body. The involvement of both cortex and brainstem in this sort of adjustment has been demonstrated (Gahéry and Massion, 1981).

The concept of postural control can also be applied to the ocular motor system. The eyes are rapidly shifted from place to place by movements called saccades, and the signal used to drive these movements specifies the velocity at which they are to occur. But when a saccade has stopped and an image of interest falls on the center of vision, the ocular motor apparatus must hold the eyes still until the next movement. The eyes must be stabilized even though the elastic tissues of the orbit attempt to return the eyes to their primary position. The stabilizing signal is produced by the cerebellum and brainstem which integrate the velocity signal to obtain a signal specifying the new rest position. For instance, if the eyes move at 100 degrees per second for 0.05 of a second, the integrator supplies a signal that keeps the eyes 100 x 0.05 = 5 degrees from the original position (Robinson, 1986).

## Low-level Coordination

The ultimate destination for both motor cortical and brainstem postural center output is the motor neurons. A major portion of the cortical and brainstem output goes, however, not to motor neurons but to the interneurons that surround them. The advantage of a direct connection of descending influences onto motor neurons is speed: some actions, such as shifting away from danger, have to be executed as quickly as possible (for example, see the LeDoux model in The Big Picture). When descending influences make their connections to motor neurons via interneurons, on the other hand, there is the advantage of using intrinsic spinal cord circuitry to coordinate muscles and modulate local reflexes.

## Driving Individual Muscles

The final link in the motor sequence is the motor unit; that is, the motor neuron and the muscle cells it innervates. Sherrington called the motor unit the final common path since all motor commands ultimately have to pass through it to have any effect. Recent evidence (De Luca and Erim, 1994) indicates that all the motor units innervating a single muscle receive a common drive signal from higher motor areas. As the drive signal increases, larger motor units are recruited and the contractile force of the muscle also increases. The provision of a single driving signal relieves the higher motor areas of the task of producing a different signal for each motor unit. This is another example of successively lower elements of the motor system breaking down the motor command into finer and finer components.

## Feedback Control

The environment in which the motor systems operate is constantly changing and often unpredictable. When you are used to driving a car without power steering, your first few turns of the steering wheel in a car with power steering are usually too vigorous. If our motor systems were purely feedforward, such errors would go uncorrected. In fact, the motor systems have evolved sophisticated feedback systems at several levels to cope with such variable loads. Feedback uses the information from a variety of receptors to compare actual movement with intended movement; a difference between the two is used to drive the motor system until the error signal is minimized. Several feedback systems — spinal, brainstem, and cortical reflexes have been described already. I describe below two other motor areas, the cerebellum and basal ganglia, that play a major role in feedback control.

## Cerebellum

The cerebellum receives a torrent of information required for its feedback task. From the motor cortex and brainstem it receives 'copies' of the motor commands descending towards the musculature. It also receives sensory information from cutaneous, muscle, vestibular, and visual receptors describing the actual movement performed. From these two types of signal the cerebellum can calculate an error signal to send back to the motor cortex and brainstem.

## Basal Ganglia

The basal ganglia (see anatomy chapter for details) receive their input from a wide area of the cortex and form a major feedback loop via the thalamus with the cortex. The projection of the basal ganglia to the premotor cortex suggests that they may play a role in the planning of movement. The motor functions of the basal ganglia were originally termed extrapyramidal since they do not contribute directly to the pyramidal (that is, corticospinal) pathway. This term has fallen out of favor, however, since the basal ganglia are so tightly linked into the cortical motor circuits. The importance of the basal ganglia in the execution of movement is illustrated by the severity of the disabilities produced by diseases affecting the basal ganglia, including Parkinson's and Huntington's disease (see chapter 11 on models of the brain in neurology).

# A DYNAMICAL SENSORY-MOTOR PERSPECTIVE

No description of sensory-motor processes should preclude a perspective of their possible ongoing dynamics, and interaction with decision-making processes. There are however large gaps in our knowledge of the sensory and motor processes of the brain. In general, our understanding of these processes is quite advanced at the receptor and effector level, but becomes steadily sketchier and more speculative as we consider more integrated central processes.

It is to be hoped that the models described in this chapter and the complementary methods and models described in other chapters, will help to provide a more integrated conceptualization of sensory-motor processes.

In this spirit, I end this chapter with a brief reference to Walter Freeman's dynamical perspective of sensory-motor processes, which he partially outlines in chapter 10b, and add here extracts from pages 75–81 in his book *Societies of Brains* (1995):

> Each sensory modality must have access to a short-term memory in a cognitive map. Each must also have access to the motor commands that move the sense organs through the work and modify its sensory input. In intentional behavior, multimodal sensory convergence takes place first, and the combined sensory input is then integrated over time and located in space. There is only one cognitive map with memory that serves this function for all of the senses, not one for each sense, and it is located in an area of association cortex called the hippocampus.
>
> The hippocampus interacts with two parts of the striatum, the septum (in humans the nucleus accumbens) having two-way connections with the hypothalamus for autonomic regulation, and the amygdaloid nucleus having two-way connections with the motor nuclei in the brainstem for regulation of the musculoskeletal system. The septum and amygdaloid work closely together (Kirkpatrick, 1994; Kalivas and Barnes, 1993). No intentional behavior is begun without prior and concomitant activation of the cardiovascular, respiratory, and endocrine systems that provide the necessary set, tuning, fuel and oxygen. This explains the sympathetic discharge upon threat of danger, whether or not physical action is engaged in. It is the output and not the input of sensory cortex that is arranged by the hippocampus. This is not filtering or template matching.
>
> The cerebellum and striatum do not set goals, initiate movements, coalesce temporal sequences of multimodal sensory input, or provide orientation to the spatial environment. These functions are performed by the limbic system. Its front door is the entorhinal cortex (Lorente de Nó, 1934), which gets input from all the sensory cortices and sends it to the hippocampus, where it is integrated over time. Hippocampal output goes back to the entorhinal cortex, which returns it to all of the sensory cortices, and updates them to expect new sensory input (Freeman, 1990a). The entorhinal cortex is the key part in human brain architecture for multisensory convergence.
>
> Most frontal lobe output goes to the limbic lobe, from which it is returned through the amygdaloid: to other parts of the striatum from which it returns through the thalamus; and to the cerebellum, from which it returns by the thalamus. The enormous feedback paths between cortex, striatum, and cerebellum indicate that the cortex uses them to prepare movements prior to execution by imagination and mental rehearsal, constituting practice with

continual additions to procedural memory by learning (Squire, 1987).

Error detection and compensation for movement induced changes in sensory input are important aspects of reafference, but transcending these, I propose (Freeman, 1990a), is its putative role in expectancy and selective attention. When an animal undertakes a search by eye movement or a sniff, it is guided not just by a prior odor or glimpse but by inputs from all of its senses, which have converged into the entorhinal cortex and combined in the hippocampus. The anatomical basis exists for returning that output to all of the sensory cortices. Updated entorhinal output may act to bias every sensory cortex in a direction appropriate for the predicted input coming from its own exteroceptor array, but in the context of recent input from all sensory arrays thus maintaining the unity of intentionality (Freeman, 1995).

## REFERENCES AND SUGGESTED FURTHER READING

Asanuma, H. (1973) Cerebral cortical control of movement. *The Physiologist*, **16**, 143–166.

Barlow, H. B. (1986) Why have multiple cortical areas? *Vision Research*, **26**, 81–90.

Darian-Smith, I., Johnson, K. O. and Dykes, R. (1973) 'Cold' fiber population innervating palmer and digital skin of the monkey: Response to cooling pulses. *Journal of Neurophysiology*, **36**, 325–346.

De Luca, C. J. & Erim, Z. (1994) Common drive of motor units in regulation of muscle force. *Trends in Neurosciences*, **17**, 299–304.

Evarts, E. V. (1968) Relation of pyramidal tract activity to force exerted during voluntary movement. *Journal of Neurophysiology*, **31**, 14–27.

Freeman, A. W. and Johnson, K. O. (1982) A model accounting for effects of vibratory amplitude on responses of cutaneous mechanoreceptors in macaque monkey. *Journal of Physiology*, **323**, 3–64.

Freeman, W. J. (1995) *Societies of Brains: A Study in the Neuroscience of Love and Hate*. Lawrence Erlbaum Associates: New Jersey.

Gahéry, Y. & Massion, J. (1981) Co-ordination between posture and movement. *Trends in Neurosciences*, **4**, 199–202.

Georgopoulos, A. P. (1986) On reaching. *Annual Review of Neuroscience*, **9**, 147–170.

Hubel, D. H. & Wiesel, T. N. (1959) Receptive fields of single neurons in the cat's striate cortex. *Journal of Physiology*, **148**, 574–591.

Matthews, P. B. C. (1991) The human stretch reflex and the motor cortex. *Trends in Neurosciences*, **14**, 87–91.

Merigan, W. H. and Maunsell, J. H. R. (1993) How parallel are the primate visual pathways? *Annual Review of Neuroscience*, **16**, 369–402.

Mishkin, M., Ungerleider, L. G. and Macko, K. A. (1983) Object vision and spatial vision: two cortical pathways. *Trends in Neurosciences*, **6**, 414–417.

Mountcastle, V. B. and Powell, T. P. S. (1959) Neural mechanisms subserving cutaneous sensibility with special reference to the role of afferent inhibition in sensory perception and discrimination. *Bulletin of the Johns Hopkins Hospital*, **105**, 201–232.

Nashner, L. M. (1982) Adaptation of human movement to altered environments. *Trends in Neurosciences*, **5**, 358–361.

Robinson, D. A. (1981) Control of eye movements. In Brooks, V. B. (Ed), *Handbook of Physiology, Section 1: The Nervous System, Volume II, Motor Control, Part 2* (pp. 1275-1320). Bethesda, Maryland: American Physiological Society.

Robinson, D. A. (1986). The systems approach to the oculomotor system. *Vision Research*, **26**, 91–99.

Roland, P. E., Larsen, B., Lassen, N. A. and Skinhøj, E. (1980) Supplementary motor area and other cortical areas in organization of voluntary movements in man. *Journal of Neurophysiology*, **43**, 118–136.

Sherrington, C. (1961) *The integrative action of the nervous system*. New Haven: Yale University Press.

Sillito, A. M., Jones, H. E., Gerstein, G. L. and West, D. C. (1994) Feature-linked synchronization of thalamic relay cell firing induced by feedback from the visual cortex. *Nature*, **369**, 479–482.

Singer, W. (1977) Control of thalamic transmission by corticofugal and ascending reticular pathways in the visual system. *Physiological Reviews*, **57**, 386–420.

Skinner, B. F. (1938) *The behavior of organisms*. New York: D. Appleton-Century Company.

# Chapter 9

# COMPUTER MODELS OF THE BRAIN

William C. Schmidt[1] and Evian Gordon[2]

[1]Department of Psychology, Dalhousie University
[2]The Brain Dynamics Centre, Westmead Hospital and Department of Psychological Medicine, University of Sydney

## INTRODUCTION

This chapter is about creating quantitative, explicit models of brain function, such as can be implemented on digital computers. Many of the models of brain function that exist today are lists of linkages or 'box and arrow' models. That is, the mechanisms discussed are communicated in a rather informal fashion and the audience has to reconstruct an image in their mind's eye of the mechanism, its workings, and its potential interactions with other components of the system.

While linkage lists and box and arrow models are adequate for describing aspects of functioning, they may obscure many of the finer details of a theory. Such imprecision may cause researchers to fail to consider alternate factors that may play an important role in the phenomenon being studied. When problems such as these start to hinder the development of a theory, delineating the theory more fully in any medium is useful explicitly to put the theory to the test. The modern digital computer provides researchers with an unprecedented tool for model building.

Breaking down a model and implementing it necessarily requires that a stand be taken on various issues related to the theory behind the model: the effort forces the modeler to adopt explicit assumptions, and these provide a way to test the model. If the assumptions of the model are falsified, an alternative more accurate approach can be sought. If the model is thoroughly inappropriate and a viable alternative exists, then such an alternative will be adopted as a successor. Through this continuing refinement process, scientists are able to gradually better approximate, understand and control the essence of the real world system that is being studied.

In the biological domain, animal models have long been used in order to explicitly test theories. A small number of factors related to the animal subject's functioning are systematically varied, and the effects on functioning are observed. This is a very effective means of studying biological systems when the effects on the system are easily observable entities, but when cognitively mediated behavior is involved, or when the effects are difficult to detect — such as in the case of a highly interconnected dynamical brain, such an approach is not always adequate or feasible.

With the recent boom in computer technology, simulation is fast becoming a standard approach to understanding all sorts of dynamical systems. Computer models can be judged on a survival of the fittest basis according to their goodness of fit with the data, their capacity to bring together explicit mechanisms and the extent of their unification rules. The next two chapters illustrate exemplar models of Brain Dynamics (including Wright *et al.*'s numerical model). This chapter outlines more traditional brain-related computer models — but in so doing serves to illustrate how brain dynamical computer models and conventional computer models can be brought together.

## TERMINOLOGY AND DISTINCTIONS

Simulation refers to a superficial resemblance (representation) of some particular aspect of the model's target phenomenon, while emulation strives to achieve equivalence in all respects.

Digital computers may be able to simulate brain processes, but it is questionable whether they can emulate them. It is very easy to fall victim to the error of treating the simulation as if it were the real thing. Simulation necessarily abstracts certain properties of the target system in order to be able to capture the essence of the system's dynamics.

A further general distinction that is important is between analogue and digital computation. Digital computation essentially means representing states of affairs in terms of logical procedures corresponding to the symbolic operations of a calculation. Digital computers carry out their computations using steps specified in such a formal code. Ultimately, the symbols of digital computation are discrete states (concatenations of binary states that are 'on or off'). In contrast, analogue computation occurs through the manipulation of continuous physical variables. We know that the human brain, being organic, operates in an analogue fashion. What we can be less sure about is whether a digital computer is capable of emulating the processes that occur in the brain. The digital computer can certainly simulate such processes, but there is always some degree of abstraction or imprecision involved in these simulations. Whether this imprecision will ultimately affect the types of processes that can be accurately modeled remains to be seen.

## COMPUTER SCIENCE APPROACHES TO MODELING MIND AND BRAIN

The computational approach to understanding the brain and mind began in the late 1950s with the creation of an independent discipline dubbed artificial intelligence[1], or AI. Strong AI[2], or that area of the field that adopted the goal of building a fully functioning intelligent automaton using information processing technology, soon discovered that the task was extraordinarily complicated. As a result, the field has developed simultane-

---

1    Not everybody agrees on the appropriateness of Artificial Intelligence as the name for the field.
2    A distinction is commonly drawn between Strong and Weak AI. In Strong AI, computers are capable of emulating human intelligence in the strictest sense, complete with all of its subtleties. Weak AI holds that computers are capable of performing acts that if performed by a human would be deemed intelligent, but does not claim that computers could in principle emulate human intelligence.

ously in a number of specialized project areas, each of which requires intelligent solutions, but not necessarily solutions that are executed in a similar fashion to the brain.

As will be discussed in the chapter on psychological models, a good research strategy to adopt in studying intelligent agents is to break their behaviors down into domain-specific components that can be more manageably investigated. AI has been forced to adopt this modular processing approach in order to restrict the amount of information that any one task has to deal with, lest the entire project fail due to the overwhelming number of factors involved. For example, machine vision alone has become a vast topic area specializing in the engineering of machines that can visually represent and interact with their environments. Within machine vision are numerous sub-domains investigating topics such as visually guided robotic control or visual object identification. Even sub-domains of visual object identification exist, with specific solutions having been devised for classes of objects that a particular computer system must specialize on identifying (for example auto parts in a computer manufacturing line).

In general, the AI of today deals with specialized, engineered solutions to specific problems, without much concern for how closely the solution matches human cognition, and with practically no concern for the role of the brain in human cognition. The human solution to a given problem domain is of interest to AI practitioners only insofar as it yields further insight into a successful automated solution. This cross-fertilizing of domains operates in the opposite direction as well. In the absence of a well-defined theory of the human behavior, computational solutions supply interesting prospective hypotheses about how brains might accomplish a given task.

## MARR'S LEVELS OF DESCRIPTION

One modular area of inquiry in which AI has managed to make significant progress is that of human and machine vision. David Marr's popular theoretical framework for approaching problem domains in order to construct computer simulations has spread throughout the cognitive sciences, and is now often used to organize the study of various cognitive phenomena.

Marr recognized that rarely can a complex system be understood merely by extrapolating from the properties of its elementary components. To investigate complex systems, Marr (1982) stressed that researchers must be prepared to adopt different kinds of explanations at different levels of description to account for the system viewed as a whole. Marr proposed that there are multiple levels of analysis for brain phenomena and that it is not sufficient merely to provide a simulation account of only one level. The three levels of interest that Marr specified are the computational theory, the representation and algorithm, and the hardware implementation.

The computational theory is the level of *what* the system does and *why* it does it. In order to realize the computational theory, a decision must be made about the representation to be used and its associated algorithm; that is, a decision must be made about *how* the computational theory is to be carried out. The most basic of Marr's descriptive levels concerns how the system is physically accomplished, or in what medium the higher levels of description are realized. Clearly the brain operates in a neural medium. The computational theory that various brain processes instantiate, however, may be implemented in other mediums, such as the digital computer.

AI researchers are interested in computational theory, as well as representation and algorithm. However, they are not interested in restricting themselves to the neural medium: rather a digital implementation is their goal. Because the representational format of neurons is unique to the neural medium, it is not of sole interest to AI practitioners. Researchers that are interested in the internal workings of the human brain, however, (such as computational neuroscientists) wish to discover how various processes are represented in neural matter. Both AI and the brain sciences share the goals of understanding similar phenomena at the level of computational theory.

## SYMBOLIC AI

Many early modeling efforts in AI did not target the brain specifically, but modeled a process that we attribute to the brain — thinking. A great variety of algorithms and representations have been created and applied to a large number of domain-specific problems. The heterogeneity of this work does not lend itself to a short review, so one branch of such modeling efforts is focussed on.

Some of the earliest AI approaches attempted to model high level cognitive phenomena, to produce automated reasoning. This work used human reasoning as a starting point. The goal was to produce streams of output similar to a human thinking aloud, and to make errors in ways similar to humans.

The work of Allen Newell and Herbert Simon helped to lay the foundations for what is now termed classical or symbolic AI. The underlying philosophy is that intelligent action is specifiable (i.e. Marr's level of computational theory) in abstraction from the implementation substrate — that is, neurons (in the case of brains). Newell and Simon stated that a physical Symbol System has the necessary and sufficient means for intelligent behavior. This entails that intelligence or thought can be constituted by symbols and operations upon them (regardless of the material substrate). Remember that Symbolic AI modelers model high level phenomena (like how we recognize objects or make decisions). The Symbolic AI approach involves the manipulation of symbols in a systematic rule driven or algorithmic fashion. And the symbols designate processes believed to be required in order to carry out intelligent behavior. Symbols may designate in several ways. For instance, a symbol may indicate that a particular action is to occur — so the computer program would proceed to the next step if a rule in the program was carried out satisfactorily. The important thing about symbols is that they represent, or stand in the place of, other things. The key to generating cognition is to transform mechanistically sets of symbols into new sets of symbols. Such transformation is thought to be similar to the way in which humans think.

Newell, in an attempt to create a more human-like thinking system, went on to propose core components of human intelligence — such as perception, memory and problem solving. Then he created a cognitive architecture known as SOAR (Newell, 1990). Newell used the term 'cognitive architecture' to refer to the base that he believed cognition relied upon. The architecture supplied the primitives from which cognition would be built, much like a computer processor supplies the basic instruction set out of which a computer program is created.

Newell's research group worked to implement a variety of cognitive reasoning tasks using their framework, adding components to the architecture that they deemed necessary.

Newell's SOAR is a sophisticated type of production system. Production systems are a collection of IF ... THEN propositions, where the left-hand side of the proposition specifies a conjunction of conditions that must hold true if actions specified by the right-hand side of the proposition are to be executed. The SOAR architecture specifies a systematic way of matching the current state of a working memory to the IF side of the productions. SOAR supplies methods to resolve conflicts should more than one production rule provide a suitable match. The repeated cycling through the productions firing appropriate productions (executing the actions specified by their THEN component) and thereby altering the internal state of a working memory takes the SOAR model on a journey through a workspace of symbols that is considered by its many protagonists to be akin to reasoning. Learning in such models is accomplished through backtracking procedures. Reasoning goals are established by the model and once such a state is met, backtracking occurs to establish the path of reasoning that led to the goal state, and the conditions leading up to the goal can be solidified, or *chunked* as knowledge, by including them as the left-hand side of new productions. Future attempts by the system to reach the same or similar goals will now result in the more efficient firing of the new productions in a single step, rather than executing a large number of productions in order to reach the same end.

In summary, the SOAR architecture provides a means of exploring problem spaces by providing production-matching procedures. It also has methods of resolving conflicts or impasses in reasoning, and learning from past experience.

The important thing about SOAR is that it is an attempt at supplying a cognitive architecture — that is, the basic components (structural and procedural) out of which cognition might be built. Faced with the product of the human cognitive architecture (namely, thinking humans) it is difficult to conceive what core components underlie it (just as when faced with the simple text output of a computer program it is difficult to conceive of what type of computer architecture produced it, or what the core instructional components being executed in that medium were responsible for the observed output). SOAR supplies a hypothesis about a set of operations and a control structure for those operations that has yielded interesting models of a wide variety of problems.

Newell and colleagues also investigated the use of parallel-distributed matching systems that used neural-like elements in order to see how such a system would alter SOAR's performance. NeuroSOAR used artificial neural network techniques to carry out some of its underlying computations and the symbol system later dealt with the outcomes of these computations in a more rigid, symbolic fashion. NeuroSOAR was able to demonstrate limited successes in several areas in which the conventional SOAR matching algorithm performed poorly.

## ARTIFICIAL NEURAL NETWORKS
### A Short History of Neural Network Research

One of the earliest approaches in AI towards modeling the brain using the digital computer was spearheaded by Warren McCulloch and Walter Pitts (1943), who argued that the digital computer was capable of simulating the human nervous system. They suggested that nervous system operations could be reduced to a collection of binary states, with each neuron roughly firing or not firing based upon incoming stimulation. Because the basis for digital computation is also a collection of binary states (bits are either on or off), McCulloch and Pitts reasoned that the digital computer is a suitable medium for implementing brain processes. Since the time of this initial proposal, researchers have recognized the oversimplification inherent in this original thesis, but this has not halted research involving the simulation of brain processes.

During the late 1960s a lot of work was done, particularly by pioneer Frank Rosenblatt, on simulating neural networks at an abstract level. This work was based on the simplest of assumptions: that neurons are highly interconnected, and have a nonlinear response to their integrated input. This is, roughly speaking, a summary of neuroscience knowledge in the late 19th century. The simulated neuron is extremely simple and looks like a caricature in the light of the vast body of current neuroscience knowledge. And yet it turns out to be a rather productive and revealing caricature. Productive because such simplification makes it possible to implement neural networks in the form of computer programs or in integrated circuits; and revealing because such simulations have been clearly successful in reproducing certain brain-like functions such as content addressable memories (the ability to recall information on the basis of an informative probe — like putting a name to a face, rather than recalling information based on the location where it has been stored), pattern completion (the ability to take a partial input and to predict what the most likely completed version of that input is, such as recovering an object identity given a partial or fuzzy view of the object), and parallel processing (the ability for multiple, spatially distributed processing units to contribute to solving a task). In addition, there are qualities to their computation that are shared by the brain such as a resistance against complete breakdown in the face of damage, and graded, continuous representations as opposed to discrete representational formats.

The original AI research on artificial neural networks, or ANNs, has spawned a large body of research across many disciplines. In physics these computational mechanisms are studied as alternative ways to implement physical models that deal with a large number of simultaneous interactions. In particular, they may be well suited for investigating the properties of dynamical systems, including the brain. In statistics the relation of these mechanisms to common statistical procedures is studied. In computer science the computational properties of these mechanisms are being investigated, as are more efficient ways to implement and apply this technology. In biology and the neural sciences, neuronal network simulation attempts to make fewer abstractions than in many of the other sciences, and as a result has as its goal the creation of the most accurate and detailed simulations possible. Because the accurate simulation of even small networks of neurons is an ambitious task, this domain is restricted to rather simple circuits compared to the complexity involved in real brains. Finally, in the brain sciences and psychology, abstract simulations are employed to gain insight into the overall workings of the brain and behavior.

For a time during the 1970s, relatively little ANN research was undertaken. This was due principally to critiques of the capabilities of simple two-layered learning networks (or perceptrons). Multi-layered perceptrons emerged using the re-invented back-propagation learning rule[3] that are potentially capable of learning any mapping

---

3   Although the problem of robust learning in multiple-layer networks had been solved as early as 1969 by researchers Bryson and Ho, the Minsky and Papert thesis overshadowed this breakthrough. It wasn't until the rediscovery of this work in the late 1980s that it's importance was realized.

from input to output. This ground breaking work by Rumelhart, Hinton, and Williams (1986) precipitated a resurgent interest in these modeling methods and this general approach to understanding the workings of cognitive phenomena has come to be popularly known as 'connectionism'.

## The Perceptron and Multi-layered Perceptrons

Perceptrons are a class of ANN which have a flow of activation in a single direction (from the input side to the output side) and which learn to associate a pattern of activation on their input nodes with a pattern of activation on their output nodes. This learning is inductive (through exposure to a number of training instances a solution is determined) and accomplished through a simple algorithm that will be described shortly.

As an example of a simple network, consider the perceptron pictured in figure 9.1A. This network consists of two input nodes and one output node. Each of these nodes, or units, is connected by a single link. Each link, or connection, has a numeric weight associated with it[4]. The specific values used for the weights can either be randomly determined when the network is constructed, or they can be hand assigned by the experimenter according to their beliefs about how the network should be constructed.

The experimenter begins training and testing by creating a set of input and output training patterns that the network will be exposed to. These input and output pairs may represent environmental objects or relations. Training would consist of presenting the network with the input and determining that it's output is based on that input. The polarity of the weights is important, as is the magnitude. Low weights will result in low activations, while large weights will result in strong activations. Negative weights will result in inhibitory effects from one node to the next. Positive weights will result in excitatory effects.

Weights play a crucial role in the network. First, they collectively determine what the outcome of processing will be for each of the units. Second, they can act as a long term storage in the network. Training a network to learn input–output associations consists of adjusting the value of the weights such that the discrepancy between the values that the network produces on its output nodes and the values that the experimenter wants the network to produce is lessened. Algorithms exist that will evenly distribute the blame for observed discrepancy across the weights, and modify them such that they are less likely collectively to produce an erroneous output after they are adjusted. It takes many applications of these learning algorithms, and hence many training examples and adjustments before a network will enhance its performance to the desired accuracy.

Astute readers might protest that there are several ways in which the ANN model does not remain faithful, even abstractly, to the target of its simulation — real neural systems. This is quite correct. For instance, while back-propagation requires the communication of an error signal from the output back to the input, there is no known mechanism by which this occurs in brains. Additionally, while training requires many examples and a great deal of computational power, humans can exhibit learning with single examples. Finally, learning using these algorithms involves an all-knowing teacher that provides information about the target of learning. Natural systems, however, learn by optimal adaptation to their environment by trial-and-error.

---

4    The terms link, connection and weight are used interchangeably.

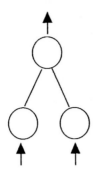

**Figure 9.1A**

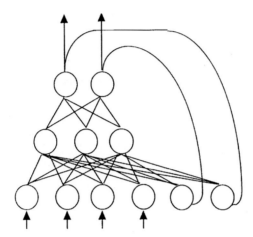

**Figure 9.1B**

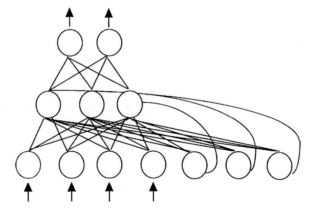

**Figure 9.1C**

**Figure 9.1A**. A simple Perceptron and some example input–output pairs.
**Figure 9.1B**. A Jordan network. Output unit activations feed back to a pair of context units which influence the network's behavior on the subsequent time slice.
**Figure 9.1C**. An Elman network. Output unit activations from hidden units feed back to as many context units. These networks learn temporal sequences.

Although there are several unrealistic (unbiological) aspects of ANN simulations, the utility of these models is in their provision of a tool that scientists can use to create better models and produce more concrete predictions than box and arrow models.

## THE CONNECTIONIST FRAMEWORK

Connectionism describes ANN techniques as they are applied to brain and cognitive phenomena. According to the connectionist approach real neural networks owe their effectiveness to the sheer number of connections and neurons that they contain, the massive distributed parallel computation that these afford, and to the adaptable strengths of synapses. Such properties can be studied with little or no reference to the detailed subtleties of factors such as membrane physiology. The name harkens back to the philosophical roots underlying this approach which holds that the power of computation lies in the way in which a system is connected, rather than within the individual units that are connected.

There are many different models and methods for weight setting and computation that can be performed with simple simulated neural elements. Connectionism is best thought of as a framework, or an approach to modeling rather than a model as such. Using the constraints of the connectionist framework, a model of any process can, in principle, be implemented.

Rumelhart (1989) notes that there are seven basic components to any connectionist system:

1   a set of processing units
2   a state of activation defined over the processing units
3   an output function for each unit that maps its state of activation into an output activation
4   a pattern of connectivity among units
5   an activation rule for combining the inputs impinging on a unit with its current state to produce a new level of activation for the unit
6   a learning rule whereby patterns of connectivity are modified by experience
7   an environment within which the system must operate.

Together, these core components are thought to be capable of simulating brain-style computation. The processing units roughly correspond to neurons or groups of neurons, hence all of the units have some tonic level of activity. Each unit is connected to others. Innervating activation is integrated separately for each unit according to the unit's specific activation rule, and then the unit's output function determines whether the unit will 'fire', thereby passing on its activation to all of the units it is connected to. Connections, between units, each have a value corresponding to the strength of association between the units that they join. Finally, all of this is immersed in a dynamic environment in which external stimuli continuously impinge on the network, resulting in network activation.

From the above outline, one can surmise that there is a large degree of freedom in the specification of individual models. A wide class of mechanisms has been investigated for each of these components, and there is no real limit to the number of forms that each component can take on. Furthermore, there are no constraints that networks have to be homogenous with respect to any one component. For instance a network could exist in which each unit has a unique activation rule, or in which each unit has a specialized output function. The only limits currently are manageability.

From this summary, we can see that there are many parameters available in the connectionist framework for the modeler to use in generating a more realistic simulation (each weight, for instance, is a free parameter, as is each variable in the algorithms used for weight adjustment). These parameters, while possibly analogous to aspects of real neural systems, provide a large number of degrees of freedom that some critics argue allows ANNs to model anything that involves a dynamical system, and the fact that they may have been neurally inspired is merely coincidence. No reference is made to our current extensive knowledge of neurotransmitters and their effects, to membrane physiology, to molecular biology, or to a wealth of other brain-related knowledge.

## Network Topologies and Learning Styles

The most common type of supervised feed-forward learning network is the multi-layer perceptron. A number of interesting variants on the basic feed-forward multilayer perceptron have been made. Jordan (1986) examined the effects of recycling the values on the output units at time t, as input to the network at time t+1 (see figure 9.1B). The extra input units, known as 'context units', are not exposed to the environment, but act to allow recurrent entry of signals to the hidden layer. Over time, the network can become sensitive to its own output from the previous time step, and adapt to change its future output based on its own behavior. Such networks have been likened to a sensory-motor system which monitors itself while executing complex actions. Another recurrent network variant was introduced by Elman (1990). Rather than recycling values from output units, Elman connected hidden units to context units which in turn fed back into the hidden unit layer (see figure 9.1C). This topology produces a network which is sensitive to the temporal dynamics of the input patterns presented. Through learning, the hidden units come to represent entire trajectories of input patterns by predicting the input pattern that will appear on the next time step.

One popular unsupervised learning method is Hebbian learning (Donald Hebb, 1949). Hebbian learning is a simple, biologically plausible form of learning in which each weight is adjusted in proportion to the correlation of activity on its input and output. Hebbian learning is simply associational, so there is no need for a 'teacher signal'. Unfortunately, Hebbian networks have yet to show great power in the types of relations they can induce.

In other unsupervised learning paradigms, the input vectors in the training set are classified into clusters of similar items and each cluster is associated by the network with a prototype vector. Such a paradigm is unsupervised because only the input side information is required. One form of clustering network based on the adaptive resonance theory of Stephen Grossberg and Gail Carpenter (1986) has strong biological motivation. Another form, the self-organizing map originally developed by Teuvo Kohonen (1992), was biologically motivated in the sense that the cluster prototypes could be organized to have a topological structure in a low dimensional space analogous to the topological mapping of features on the cortex.

The diversity of these new network configurations seem likely rapidly to reshape the Classical versus Connectionist debate. The classical approach aims to model cognition: that is high level reasoning, problem solving, and the types of tasks that we would consider to require intelligence if a human were executing them. These models can take any of a wide variety of forms, all of which have in common that they use symbols to designate the entities to which they are referring. The goal of classical models is to systematically manipulate these symbols in a fashion that is similar to human cognition, and there are innumerable ways of doing this. Connectionists, on the other hand, are primarily concerned with making models of tasks which have some degree of resemblance to the way that human brains might operate. Again, there are innumerable ways that the elementary constituents of connectionism can be put together to form a model. The true yardstick of utility is the degree to which any of these models assists in discovery and the stimulation of more refined research questions. On this measure, both approaches are likely to remain fruitful, and both show potential for integration of diverse intellectual input into these models.

## LEVELS OF BRAIN MODELING (BIOLOGICAL COMPUTER MODELS)

Just as AI has been forced to adopt a modular processing approach in order to restrict the amount of information that any one task has to accommodate, so too have the brain sciences. This means, for example, modeling auditory processing independently of visual or olfactory processing. That is an obvious kind of partitioning. In addition, there is a need to separate the problem into levels, which has provided numerous successful outcomes in the broader reductionist strategy in the neurosciences.

Perception, language, memory and problem solving are all indisputably high level phenomena, and hard to relate to synapses and action potentials, but it is an assumed doctrine in the sciences that high level phenomena do have an explanation in terms of lower levels. Put another way, cognition is an emergent phenomenon of living neural networks seemingly unrelated to lower levels of description. But a relationship does exist with neurons: a mysterious relationship given our current state of knowledge, but certainly not a mystical one. We have deliberately in this book made the notion of reductionism in the brain sciences concrete: into a neuronal level, a network level and a whole brain level. The first is rather well understood, but with the others we are still feeling the way. We have also emphasized that different organizational mechanisms may characterize different levels of function (different rules of function at different scales).

We shall now consider, in turn, modeling approaches at the neuronal, the network, and the whole brain levels (in chapters 10a and 10b). This is a somewhat arbitrary and even controversial division, but one that serves the purpose of demonstrating the wide range of brain modeling that is currently undertaken. Neuronal modeling attempts to incorporate as much real world knowledge as possible into the simulation, in order to further study neural response at the microscopic level. Neural network models come in two varieties (as previously outlined): those which attempt at simulating realistic networks (emulation), and those that abstract away from the possibly unknown specifics in an effort to gain insight into specific problem domains (simulation). At this stage of our knowledge, whole brain models make numerous assumptions, but their goal is to gain insight into overall brain organization at a macroscopic level.

## Neuronal modeling

Neuronal modeling is explicitly concerned with simulating the nervous system's microscopic function to a high level of detail, without primary consideration towards obtaining higher level effects. However, these models represent the foundation on which higher level models ultimately rest. As discussed elsewhere in this book, the picture of neurons that has emerged in the last 100 years is of cells that are physiologically similar to each other, but are diverse in size and shape. Some neurons are small, with few dendrites and an axon less than a millimeter long. Others are large: neurons controlling our legs can be a meter in length, making them the largest cells in the body.

Current neuronal models are extremely successful at representing dendritic polarization changes, and the propagation of action potentials. Thanks to the ability of computers to handle both complicated equations and complicated constraints, modeling may be carried out that is detailed in both the spatial and the temporal domain. For example, one can model dendritic depolarization in the dendritic tree, and as well see the time course of activity in great detail.

A model by Koch and Segev (1989), for example, incorporates 11 different types of ion channel, and draws extensively on molecular biology as well as on microanatomy. The success of neuronal level models is gratifying, but also disappointing to anyone who expected the full, impressive range of human brain function to be, in some sense, explained. In essence, this is because overall brain function is governed by all *interactions* between the component parts, rather than by the component parts themselves. Furthermore, there are technological limitations that prevent neuronal models from scaling up to the level that one would need for modeling even the simplest of whole brain mechanisms.

A key fact about the brain that is emphasized in this book is its high degree of interconnectedness. There is a combinatorial explosion of possible interactions at the network and whole brain level. Models are the only way to explore the essence of these interactions.

## Network modeling

Network modeling addresses the issue of higher level interactions, and is the archetypal application of the connectionist approach. As discussed earlier, consistent with the belief that it is the connections that matter, the 'neurons' in such models are quite simple, and the main focus is on finding the appropriate set of interconnections to achieve a given goal. In the engineering context, the goal is usually that of classification: given an input pattern, which group does it belong to? For example, given the spectrum of a short period of speech, what are the corresponding phonemes? Or given an image of a face, is it one of the faces 'known' by the network?

In the emerging field of 'Computational Neuroscience' (Churchland and Sejnowski, 1996), the challenge is to model not just the output, but also the real inner workings of neural networks. This is in some ways an additional burden for the modeler, but, from another perspective, it is made a little easier because we can get guidance from experimental sources. Moreover it promises great insights: for it is an opportunity to learn how Nature solves computational problems of a difficulty that still defies science. Despite all the advances in these fields, we are still quite unable to reproduce activities that any animal carries out 'without thinking'. However a start has been made in understanding the manner in which neural networks achieve functionally useful ends.

The increasing emphasis on brain modeling that incorporates biologically faithful parameters, and sequences of higher level function explained by lower level function, may be the key to bridging the gap between the neuron and the conscious brain. *The Computational Brain* (Churchland and Sejnowski, 1996) and *Neuroinformatics* (Koslow and Huerta, 1997) distil many of these efforts and point the way forward in this regard.

Yet even with simplified neural structure and learning mechanisms, it is still impractical to model physiological networks or ANNs of over about 1000 neurons. That is why there is another level of modeling that tries to explicate the collective behavior of many thousands of neurons and even aspects of whole brain function. This emerging field of whole brain modeling is addressed in the next two chapters on 'Brain Dynamics'.

## REFERENCES AND SUGGESTED FURTHER READING

Amit, D. J (1998) *Modeling Brain Function*. The world of attractor neural networks.

Churchland, P. and Sejnowski, T. (1996) *The Computational Brain*. Cambridge, MA, MIT Press.

Copeland, J. (1993) *Artificial Intelligence — A Philosophical Introduction*. Cambridge, MA, Blackwell.

Elman, J. L. (1990) Finding structure in time. *Cognitive Science*, **14**, 179–211.

Fodor, J.A. (1981) The mind-body problem. *Scientific American*, **244**, 114–25.

Fodor, J. A., and Pylyshyn, Z. W. (1988) Connectionism and cognitive architecture. *Cognition*, **28**, 3–71

Franklin, S (1998) *Artificial Minds*. The MIT Press.

Grossberg, S., and Carpenter, G. A. (1986) Adaptive resonance theory: Stable self-organization of neural recognition codes in response to arbitrary lists of input patterns. In *Proceedings of the Cognitive Science Society*, Hillsdale NJ., Erlbaum.

Haugeland, J. (1985) *Artificial Intelligence — The Very Idea*. Boston, MA, MIT Press.

Hebb, D. O. (1949) *The Organization of Behavior*. New York, Wiley & Sons.

Hertz, J., Krogh, A. and Palmer, R. G. (1991) *Introduction to the theory of neural computation*. New York: Addison-Wesley.

Jordan, M. I. (1986) Attractor dynamics and parallelism in a connectionist sequential machine. In *Proceedings of the Eighth Annual Conference of the Cognitive Science Society*, Hillsdale NJ., Erlbaum.

Koch, C. and Segev, I. (1989) *Methods in Neuronal Modeling*. Cambridge, Mass., MIT Press.

Kohonen, T. (1992) Self-organized formation of topologically correct feature maps. *Biological Cybernetics*, **43**, 59–69.

Koslow S. H and Huerta M. F (1997) *Neuroinformatics. An overview of the Human Brain Project*. Lawrence Earlbaum Associates. New Jersey.

Marr, D. (1982) *Vision: A Computational Investigation into the Human Representation and Processing of Visual Information*. San Francisco, W. H. Freeman.

McCulloch, W., and Pitts, W. (1943). A logical calculus of the ideas immanent in nervous activity. *Bulletin of Mathematical Biophysics*, **5**, 115–33.

Minsky, M., and Papert, S. (1968) *Perceptrons*. Cambridge, Mass. MIT Press.

Newell, A. (1990) *Unified Theories of Cognition*. Harvard University Press. Cambridge, Massachusetts.

Rumelhart, D. E., Hinton, G. E. and Williams, R. J. (1986) Learning internal representations by error propagation. In D. E. Rumelhart and J. L. McClelland (eds.), *Parallel Distributed Processing: Explorations in the Microstructure of Cognition*. Vol. I: Foundations., chapter 8. Cambridge, MA, Bradford Books/MIT Press.

Rumelhart, D. E. (1989) The architecture of mind: A connectionist approach. In Posner, M.I. (Ed.), *Foundations of Cognitive Science*. Cambridge, MA, MIT Press/Bradford Books.

Russel, S. and Norvig, P. (1995) *Artificial Intelligence — a modern approach*. Englewood Cliffs, NJ, Prentice Hall.

# BRAIN DYNAMICS:
# MODELING THE WHOLE BRAIN IN ACTION

Jim Wright

Brain Dynamics Laboratory, Mental Health Research Institute of Victoria

## INTRODUCTION

The brain is a system, and has evolved only because its dynamic properties and its structured connections to and from the outside world are innately configured and continuously reconfigured so as to give rise to adaptive behavior, aiding the organism's survival. This biological truism raises major questions about how the co-operation of billions of neurons can achieve such a feat. Brain dynamics attempts to address these questions. The field of research is relatively new, and draws upon electrophysiology, and neural network theory. Its particular interests can be distinguished from its parent disciplines in a number of ways.

Firstly, brain dynamics is concerned with the brain as a whole, rather than analysis of its microscopic detail. Yet attempts are made to approximate the character of real neurons with more veracity than is generally needed for neural network applications within computer science.

Secondly, it has grown around study of the electroencephalogram (EEG); that is, the study of the ongoing electrical fields emitted by the working brain. The EEG remains the best and cheapest way to observe the whole brain in action, on a time scale from milliseconds to many minutes — the time-scales of mental events. This wide time-window makes the EEG an important complement to other brain scanning techniques such as PET and fMRI which resolve events in space with marked superiority to EEG, but at the cost of time-resolution (see chapter 15 Brain Imaging Technologies).

The guiding notion is that if we can achieve understanding of the dynamic rules which underlie the origin of the EEG, this will prove a high road to understanding the dynamics of the brain generally and hence insight into how the brain gives rise to adaptive behavior. Brain dynamics is thus converse and complementary in its approach to attempts made to achieve artificial intelligence with general-purpose computers.

The shared intellectual ancestry of brain dynamics with these other disciplines makes it a part of physics as well as biology. To the distaste of many biologists, mathematical procedures and methods similar to those applied to molecular interactions, the weather, the ocean waves, and the like must be used to grasp the way billions of neurons interact. To the distaste of physicists, it is very difficult to reduce the properties of neurons to sufficiently simple and quantified elementary units for study.

In this chapter we shall try to give the gist of the powerful mathematical tools which physics and computer science can offer, as well as showing how they are applied in mathematical models of brain organization.

We will also try to reach ahead of existing models, to show how an understanding of the dynamics of the brain may interlock with theories concerned with information storage and general computation, so as to lead us toward more powerful future theories.

We must begin by some further consideration of the EEG, since this is the objective measure upon which the theoretical considerations are built.

## SOME BASIC CONCEPTS RELEVANT TO BRAIN DYNAMICS
### The Origin of the Electroencephalogram (EEG)

The term EEG is usually confined to the recording of electrical fluctuations from the scalp. Recordings made directly from the brain' surface are called electrocorticograms (ECoGs). EEG and ECoG are actually differential voltages between the point of recording and some remote site on the subject's body. The head and the entire body form a conductive medium for the flow of electrical currents. This medium has electrical resistance, which varies from site to site in a complicated way. Bioelectrical voltages arising at all parts of the brain act as the sources of current flows. The current flows generated sum together at any site according to the total electrical resistance along all the possible paths of current flow between the sources and the site. The recorded voltage at the electrode thus reflects the summed current from all sources, and fluctuates over time as the bioelectrical voltages fluctuate.

The diffuse nature of the current flows, and the size of the recording electrodes, conspire to blur the picture of brain activity thus obtained. Ideally, the EEG from each recording electrode would represent only the activity from brain cells immediately adjacent to the recording site. The blurring effect changes both the frequency content observed in the signal, and clarity of the 'image' of electrical activity on the brain's surface. The change in the signal observed, from the idealized source activity, is termed temporal and spatial low-pass filtering, because slow events and spread-out fields of activity are selectively represented in the recording. The larger and more remote the recording electrode from the sources in the brain, the greater these filtering effects usually are.

With sufficiently small electrodes situated within the cortex itself, it can be shown that the primary sources are electrical fluctuations arising within the dendrites of the cortical cells. The local voltage fluctuations in dendrites within a small zone of neurons summed by the small intracortical electrode is called the local field potential (LFP), and it is the summing and filtering of LFP from all over the cortex which gives rise to the EEG. The LFP ranges in frequency from less than 1 Hz (cycle per second) to around 100 Hz. LFP does not directly reflect the action potentials (pulses) of the neurons' activity. The pulses transmitted via the axons involve events of much higher frequency content, and even though these act as comparatively large voltage sources, they are excluded from the EEG recording by the spatial and temporal filtering.

The variations in the membrane potentials of the cells giving rise to LFP are several millivolts in amplitude, and have the same frequency components as are present in the EEG. The underlying fluctuations in potential within the dendritic membranes are called neuronal waves. Neuronal waves arise in turn from the continuous bombardment of the dendritic membranes by synaptic action, as action potentials arriving from other cells' release of neurotransmitter chemicals. The neurotransmitters change the conductivity of the dendritic membrane to ions (sodium, potassium, chloride) bathing the dendritic membrane. As the ions flux back and forth across the dendritic membranes, the LFP is created.

All cortical neurons do not contribute equally to the EEG. The cells which are the principal source of EEG are the excitatory (pyramidal) cells in layers III and V of the cortex, which are so orientated that they generate greater net currents at the cortical surface.

The problem of converting EEG signals blurred by the low-pass filtering into a more accurate picture of the original distribution of bioelectrical potentials in the brain is known as the 'inverse problem'. Computer techniques which allow for the distorting effects have been developed. These 'deconvolve' the recorded signals to a form more closely resembling the distribution of sources at the brain's surface. As yet none of these techniques is wholly satisfactory, but fortunately this has not greatly impeded the development of dynamical theory for neuron interactions.

## The relation of the EEG to cognitive events

The EEG has often been dismissed as an epiphenomenon having little if any strong relationship to more fundamental events in the brain whatever these are. The exchange of action potentials between neurons is readily identified as the bona fide medium of information exchange between the neurons, whereas it seems unlikely that the current flow of the EEG itself plays any significant role in the dynamics of the cells' interactions. This criticism rather misses the point that action potentials and dendritic potentials are but two faces of a single coin, as information flowing between cortical cells has to pass through both dendrites and axons. The state of the dendrites on masse is observed via the EEG, and there is no need for the EEG itself to act as a direct medium of information exchange for its evidence to be admissible as a guide to cortical information processing. Thus action potentials and EEG recordings can complement each other, and there is good reason to expect the EEG to act as a mirror of cognitive events.

A more sophisticated reservation is that the EEG cannot contain very much information about the brain's overall state since the frequency content is limited and the spatial resolution of the EEG is poor. This certainly limits the technique — but it leaves an open question whether the information that the EEG does convey is of critical functional importance concerning the brain's state. To answer this, we need to define what is critical about the overall dynamics of the brain, and show whether this is reflected in the EEG.

A further problem in assessing the EEG's usefulness as a witness to matters of importance in the brain's state, concerns controversy over whether information being transmitted between neurons by action potentials is 'frequency coded' or 'population coded'. Roughly: how much does the exact pattern and timing of individual action potentials (frequency coding) determine the transmission of information in the cortex, or is it just the statistical properties of action potentials in many cells (population coding) which constitutes the signal? If frequency coding were vital, the EEG would be more limited as a pathway of enquiry into the brain. Furthermore, any models or theories of the brain's workings would have to represent the coding of individual neuron pulses. Brain dynamics takes the view that it is cell population properties which form the proper target, for preliminary models at least.

Having noted these reservations, the following features are some of the most important which support the impression that the EEG is a 'cognitive mirror', even if a rather indistinct one:

1)  The EEG is an indicator of wakefulness. During intense alertness the EEG enters a state called 'desynchronization' in which the power spectrum (the amount of activity at each frequency) has a characteristic form, loosely called '1/f', because the power (squared amplitude) of signals at any frequency is roughly inversely proportional to the frequency.

    During restful but moderately alert states, a powerful peak of activity is present between 8 and 14 Hz — the alpha rhythm. Other frequency bands — the delta (1–3 Hz), theta (4–7 Hz), beta ( 15–25 Hz) and gamma (30–80 Hz) show activity that richly fluctuates during cognitive activity. Of all these frequencies it is activity in the gamma band, and especially around 40 Hz, which is most closely linked to action potential fluctuations. Only in the gamma range does EEG signal arise from LFP in small concentrated zones of the cortex, and activity in the gamma band is associated with higher firing rates of action potentials. At lower frequencies EEG activity arises over very large domains of cortex, and reflects the low background rates of action potentials (5–10 per second) which pertain for neurons in less excited states.

    Other EEG changes characterize sleep, and these include specific changes typical of both deep sleep and dreaming.

2)  The event-related potential (ERP) is the signal-locked events in the EEG after a discrete stimulus or event — from about 100 milliseconds after the event to a second later. Components of the ERP are exquisitely sensitive to all aspects of the attentional set — the subject's expectations, as well as different aspects of stimulus evaluation and response selection.

3)  Topographic analyses (that is, pictures of the variation of electrical pattern, obtained from many electrodes) reveal detailed patterns of electrical activity, showing the sequence in which different brain regions are orchestrated to undertake a precisely defined action. For example, a particular visual stimulus, which is the cue for a hand movement, causes a pattern of activity moving from visual to motor areas, among many other associated changes in the brain's state.

4)  Topography can be used in certain circumstances to distinguish individuals with specific disabilities, such as ADHD, dyslexia and schizophrenia.

5)    Patterns of EEG activity in the olfactory bulb reflect not only the odors presented, but the subject's appraisal of the significance of the odor.

6)    The EEG and action potentials are correlated at frequencies near 40 Hz, as mentioned above, but additionally, action potentials can be instantaneously correlated at distances of several millimeters, and EEG fields themselves over distances of centimeters, further suggesting that the EEG is correlated with both local and global signal processing. This we will address further in the next section.

To sum up this section: findings indicate that the EEG can reflect large trends in the flow of information in the brain. The image of cognition obtained via the EEG could probably never be complete, and we do not know how much of the total information being processed in the brain at a given time can be extracted from the EEG.

## Synchronous oscillations in the brain

The fields of brain activity identified by the EEG have recently been the subject of renewed interest since their association with synchronous oscillation was discovered. Individual cells in the cerebral cortex respond selectively to different aspects of a single stimulus. Some cells respond to the color of the stimulus, some to features of the shape, for instance. It has been shown by Singer and Gray, Eckhorn and others, that when two sites on the cortex are concurrently responding to separable stimulus properties such as color and shape then the active cells enter into 'synchronous oscillation'. That is, pulse trains and LFP activity at the two sites all become highly cross-correlated. Within a few milliseconds these cross-correlations are apparent without any lag between the two sites. The same is generally true for all kinds of events in the brain which go to form aspects of the perception of a single stimulus object — but synchrony does not emerge if the two stimuli have no perceptual relation to each other. An example clearly indicating the need for the stimuli to form part of a perceptual whole is provided by experiments on cats which have developed a squint because of poor muscular control of one eye. This disability leads to a failure of stereoscopic vision. Thus the two sites on the visual cortex, one from each eye, which represent the same point in visual space, do not enter into synchrony in animals which have impaired binocular vision, yet do so in normal animals.

The synchrony phenomenon was predicted by von der Malsburg, a brain theorist who was seeking a way to solve the 'binding problem'. During analysis of the organization of visual and other sensory processes, it was recognized that as signals progress up the visual pathway, individual neurons deeper and deeper in the hierarchy of cells respond selectively to different aspects of the stimulus properties, as noted above. But at higher hierarchical levels, cells also respond too increasingly complicated combinations of stimulus features. These cells may be said to bind stimulus properties. It was surmised that if this process continued, somewhere high in the brain there ought to be neurons that responded to stimuli as complicated as one's grandmother — hence the term 'grandmother cells' was applied to these theoretical entities. Calculations soon revealed that there could not be enough neurons in the brain to subserve this representation by binding on a cell-by-cell basis. Malsburg recognized that the concurrent activity of many different combinations of cells would be required to enable all the possible separate perceptual stimuli to be bound and represented in consistent groups. The phenomenon of synchronous oscillation filled that need.

But how did the synchronous activity come about? And how does the brain learn to store the relations between areas which enter synchrony to thus represent a single object? How are the patterns of synchrony assembled and disassembled? These are questions which can be considered only within the context of the dynamics of neuronal interactions: to which we can turn after considering dynamics more generally.

## Types of Dynamic Interaction in Complex Systems

It is now widely recognized that not all events in the physical world follow completely predictable courses. The earlier history of physics was concerned with the use of predictable events, selected, with great insight, precisely because they could guide insights into basic laws. For instance, the work of Isaac Newton enabled the calculation of the orbit of a single planet around the sun from the use of first principles. Newton's beautifully simple laws encounter difficulty when they are applied in more complicated situations. When many bodies interact, the exact solution of their future behaviors is often unpredictable in detail in the longer term.

The problems of finding complete rules for prediction in complicated systems are made worse when the interactions of the component units do not take place according to some smoothly graded effect, such as gravity, but involve processes which exhibit sharp discontinuities. The interaction of neurons is precisely such a sharply discontinuous process, since each cell either fires off an action potential in response to input, or it does not do so, with a definite boundary of transition. Thus small differences in the inputs to a specific cell can have totally different effects upon its output. This small difference could result in a chain reaction of differences in the subsequent states of all the neurons to which it is connected. As all neurons are connected to large numbers of others, we can readily imagine this making vast differences to the exact pattern of firing of cells — their 'frequency coding', that is.

However, it does not follow that this extreme sensitivity to 'initial conditions' (whether the first cell did or did not fire) will always have a very large effect on the average firing rates of cells over periods of time much longer than a single action potential. The population coding of the cell firings might be little affected, even if the frequency coding is drastically changed over any short interval. Yet we can also imagine certain delicately balanced, unstable conditions in the brain, in which both the temporal coding and the population coding of cells could be drastically changed by events in only a few cells.

How and when does the brain utilize these dynamic complexities so as to operate both flexibly and reliably? Advances in the theory of dynamics can be used to address this problem.

Defining behaviors in very complicated systems is enabled by the use of ideas originated in the mathematical subject called topology. This subject deals in geometric relations which are independent of deformations, so long as adjacency relations of neighboring points on a surface or volume are preserved. Leaving out the features which can be accounted for by deformation, such as the difference between an egg and a sphere, enables other properties to be isolated. For maximum generality, topology deals with kinds of imagined spaces which can have more dimensions than the three dimensions of conventional space. This introduces an important concept — that of a *phase space* or *state space*.

The simplest state space is the system of axes as used in coordinate geometry or geography for that matter, having two axes at right angles in two dimensions. State space is made n-dimensional by imagining that there are n axes, each at right angles to each and every other. Each axis is used to measure the state at a given instant, of one crucial property of one elementary unit, of the system studied. For our purposes, it could be the depolarization of the dendritic membranes of one of the neurons as an axis, and the rate at which this depolarization was changing as another.

With n axes, the state of a neural system of n/2 neurons can be thus represented at an instant, by the values on the n axes. An equivalent way to state this, is that the instantaneous state of the neurons is an n-vector. For a whole brain, n would be billions, as there are billions of neurons, and if we wished to describe more details than just the depolarization of the cells we would then need a space of higher dimension still. If the brain's state has to be frequency coded to be realistic, then we are in great practical trouble. But if population codes will suffice, then we could represent the brain's overall state in a state space of lower dimension. We can say that the extent of population coding is reflected in the dimension of the necessary state-space

The movement of the n-vector within the state space over time, as events occur in the system, is called a trajectory, and adjacent trajectories form a 'flow'. For any position on a trajectory we can say we understand the physics of the system, if we can write the rules which determine the transition to the next point. The rules are prescribed by a system of equations called state-transition equations, and the physics applicable to interactions among any two elements in the system usually enables these equations to be written out so as to deal with events in n/2 elements, in n dimensions. Where n is not too ridiculously large, the equations can be used to compute the state of the system a small time interval after any given state, with appropriate allowance for the effects of fresh inputs to the system.

Once we have some idea of the appropriate reduced dimension of the state-space, the form of the transition equations, and the values of any constant terms (parameters) in the equations, then we can move on to consider the stable and unstable domains within the system's state-space. To achieve this, the flows of complicated systems are often dealt with by defining the properties of regions (hypervolumes) in the state-space, which are called attractors. There can also be hypervolumes called repellors, which are the opposite of attractors, and saddle-points with a combination of attractor and repellor features depending on the approach direction of a trajectory. There are even more complicated forms possible in higher dimensions, but the essential quality of all, is the way flows are guided into or out of the particular hypervolume.

Attractors, as the name implies, are zones in multidimensional space about which the trajectories of the system are attracted. That is, the flows tend to roll up and remain in the area of an attractor, like loops or knots in a piece of string, except for trajectories describing states so energetic that the trajectory escapes from the attractor. Each attractor represents a stable state for the system, and if we understand the structure of its attractors, then we know how it will perform from given initial conditions.

Attractors fall into three essential types.

(i)     Point attractors
(ii)    Limit cycle attractors
(iii)   Strange attractors, also called chaotic attractors.

The first of these is the simplest type. In a point attractor flows go to a point in the hypervolume and stop there. Limit cycles attractors have the added feature that motion does not cease in the system, but settles into a set of cyclical behaviors. Both these types of attractor share the property that flows from any starting point inside the attractor converge toward the static end-point or the limit cycle. That is, the ultimate behavior of the system is rather insensitive to the system's initial conditions. So long as a trajectory starts within a basin of attraction, and is not too energetic, it is irrelevant where it starts, if all you are interested in is where it finishes. In strange attractors this tendency to convergence of flows independent of their exact starting point is not present. Although the flows from any initial point proceed to swoop about in the attractor hypervolume, trajectories which are arbitrarily close together at first tend to diverge. If this diverging tendency is present for all initial points in the volume of attraction, then the trajectories are called chaotic.

These differences between types of attractor do not matter in a larger sense. What does matter is the structure of the boundaries between attractors — the unstable domains, including repellors, which define which way the system will tip, when it reaches any boundary of an attractor.

The limited sorts of attractor possible, and the fact that the global dynamics of a system can be defined in terms of zones of attraction rather than by detailed analysis of motion and state within attractor basins, leads us to a great simplification. We need not consider it our goal to account for the sequential state of all the neurons in the brain. *If we can find the structure of the brain's state space, we have got a strong guide to the crucial aspects of the overall dynamics of the brain.*

Finally on this topic, it may be useful to point out that movement in chaotic trajectories is not the same as random movement, or noise. By noise is meant a process completely random from instant to instant. Randomness may occur if some outside process is being imposed on a system's movement in its state-space. In all deterministic systems including chaotic systems, each successive state is followed entirely lawfully by its successor — although (within a specific chaotic attractor) over time a state is reached which is essentially random with respect to the initial condition.

Brain dynamical models sometimes depend on representing inputs to the cortex as random processes, usually defined as 'white noise'. A white noise is a random signal, with equal amplitudes at all frequencies over a sufficient time.

## Attractor Neural Networks

The state-space approach lends itself readily to the analysis of the behavior of simulated neural networks, and particularly to systems called attractor neural networks. This topic has been developed within computer theory, largely for practical purposes ranging far outside applications to brain theory.

An attractor neural network is a system of coupled elements designed to be like simplified real neurons. The set of all possible initial conditions (or inputs) for the network can be divided into subsets, according to which of the network's attractors the network state settles into from a given initial condition. The particular attractor into which the flow terminates then may be considered a 'decision' by the network about the initial conditions. Thus, appropriately coupled networks can act to classify fresh inputs into categories. The categories can be simple, or as complex as the recognition of particular individuals' handwriting etc.

Do sets of neurons operate under similar rules? The answer seems to be yes, at the abstract level of attractor dynamics, even though real neurons are much more complicated than their simple computer equivalents, and likewise, real neurons must have much more complicated attractors than their computer cousins. Another major question which has also been partially answered from considerations of simplified computer networks, is related to the learning rules which can determine which connections are 'appropriate' for a particular type of categorization, and how this is related to the type of dynamics in each attractor. This is considered in a following section.

## Linear versus nonlinear systems

One more aspect of the mathematics involved in brain modeling needs some general comment. This is the distinction between linear and nonlinear systems. Linear systems can be described by equations which preserve the property of additivity of crucial system variables, while nonlinear systems do not exhibit such additivity. By this is meant that if a linear system reacts to an input a, with a reaction x, and to b with y, then it reacts to (a + b) with (x + y). Linear systems are fully analyzable by formal mathematical techniques, whereas nonlinear systems can only sometimes be so handled. Because of the tractability of linear systems, one aspect of brain dynamics modeling has been a search for circumstances in which linear mathematical models of the dynamics could be applied, despite the extreme nonlinearity of the response of individual neurons.

## Cerebral anatomy, subcortical control and the EEG

A theory of the EEG, let alone of the brain's dynamics, needs to consider the specific organization of the brain's anatomy. Important aspects which require incorporation in some manner, however approximate, include:

(i)   *The anatomy of the cortex.* No unit of cortical anatomy is perfectly modular, and allowance must be made in the design of a simulated cortex, for the intertwined connections of the millions of cells — both with regard to the apparently random pattern of some of the connections, and the non-random nature of many other connections. There needs also to be regard for the cortex's organization in depth, its inhibitory and excitatory cell types, and the range of axonal connections, which range from fractions of a millimeter to many centimeters. There appears to be need also to allow for the development of regional specializations of function within the cortex: from minicolumns through macrocolumns, between the different sensory and motor areas, and at all the hierarchical orders of connected areas in between.

(ii)  *Subcortical/cortical interactions.* Cortical afferents form a numerically small part of the connectivity of the cortex, but are functionally vital. These include the specific nuclei of the thalamus, the basal ganglia, the hippocampus and related limbic circuitry. All of these are situated so as to act as to-and-fro relay systems in interaction with the cortex.

Of principal concern is the reticular activating system — a complex and diffuse system projecting from the brain stem to the cortex, which provides both excitatory and inhibitory input. This subsystem is itself under control from the cortex, as well as from collaterals of sensory pathways. The activating system includes many cell types utilizing a whole range of neurotransmitters, and is crucially concerned with the maintenance of the waking state, desynchronization of the EEG, and the direction of attention and governance of motivation.

A large body of research has identified rhythmic EEG processes with rhythmic driving from subcortical sites, but this appears to provide only a contribution to the EEG's rhythmic sources, and we must also look within the cortex itself for the origin of the rhythms.

These requirements are too large to be immediately tractable. Existing brain dynamical models have all been developed so as to attempt to represent the essential aspects of the brain's systems and circuits without being drowned in the complexity of the reality.

## SOME SPECIFIC MODELS

We can now turn to some specific theories of brain dynamic interactions and the EEG. The examples chosen are intended to give the reader a sense of the way that principles from the theory of dynamics have and are influencing the formulation of models for the brain, rather than providing a complete survey of the field. It is useful to see the models as partial theories working towards a more complete description.

### Wilson and Cowan

The work of these investigators acted as an important link between neural network modeling, as it developed in the hands of computer scientists, and EEG theories. They were among the first to apply continuum models to neural interactions. Continuum modeling uses smooth average properties of populations of cells, rather than giving an account of individual neurons as is done in neural network theory. Their work did not deal with the EEG as such. Their principal concern was certain phenomena of vision. Many of the properties of their model were carried over, explicitly or implicitly, to later theories concerned with the EEG. In their 1973 paper:

(a)   They dealt with a mixture of inhibitory and excitatory cell types, which were organized in aggregates rather than being considered a cell at a time.

(b)   They considered each cell as having a threshold potential, such that above a certain strength of input the cell would become activated and begin firing action potentials. The distribution of the thresholds in the population of cells was according to a sigmoid (s-shaped) curve relating LFP to number of cells above threshold.

(c)   They considered the cells as homogeneously distributed in a plane to form an ensemble. Cells at different distances interacted by reciprocal lateral connections, with the excitatory to inhibitory connections having a longer range than excitatory to excitatory connections.

(d) They assumed that no overall state of excitation could be maintained without external input.

They were then able to formulate equations of state for the ensemble. They adjusted certain values in these equations representing strength of couplings (the system parameters) to reproduce conditions of the structure of various parts of the brain, and then explored the attractor classes of the system.

Such ensembles of cells exhibited complicated oscillatory patterns, which included limit cycles, and sudden changes of frequency, which they called frequency demultiplication. The latter effect may be thought of as a transition between attractors in state–space.

While their findings offered an impressive account of cooperative interactions among cells, with a good qualitative match to visual findings, the experimental data of the day did not permit the model's exhaustive testing. Most but not all of their working assumptions carry over to later EEG models. The assumption that excitatory to inhibitory interactions are of longer range than excitatory-excitatory interactions is not true in the cortex, yet this property was essential to the stable performance of their model. A surround of inhibitory action was required to control the wildfire spread of excitation from any excited focal site. The mean firing rate for cortical cells is about ten impulses (spikes) per second, and their maximum possible rate about two hundred spikes per second. Therefore most cells are well below threshold most of the time, and this was not explicit in their model.

## van Rotterdam and Lopes da Silva (the Amsterdam model)

In 1982 these workers provided a detailed electrocortical model linking neural networks to the neocortical EEG with the specific intent of explaining the alpha rhythm. Of special importance in the form of this model was its emphasis on using physiological and anatomical values for various constants in the theory, and the prediction of particular wave properties which were compared with the results of experiment.

They were forced to come to terms with the fact that some of the crucial anatomical and physiological measurements needed to specify their model were not available. The lack of these crucial measurements meant that they had to introduce 'free' parameters. The 'free' parameters were cleverly lumped mathematically so they were reduced to a single number, thus avoiding the need to manipulate parameters in an unrestrained way.

Two interacting populations of cells were again considered: the excitatory pyramidal cells and the inhibitory interneurons. Each were characterized by transfer functions corresponding to the time course of excitatory and inhibitory post-synaptic potentials. (Transfer functions are linear mathematical expressions describing how inputs at different frequencies are transformed). Further functions were used to characterize the nonlinear process by which post-synaptic potentials are transformed into output pulses. The difficulty of specification of the nonlinearity was disposed of by assuming that over a short range of deviation of the density of post-synaptic potentials the nonlinear functions were effectively linear. Strengths of coupling between the cell populations reflecting the densities of synapses were introduced as constants, which were then manipulated as free parameters since they were not known with accuracy.

Paired excitatory and inhibitory populations were then arranged in a chain, connected together at ranges corresponding to the actual anatomical extent of local cortical fibers. The whole system was then considered driven by a white noise. By algebraic manipulation of the state transition equations, the free parameters were conveniently lumped so they became a single parameter. It was then shown that with a particular choice of this parameter, the chain of neurons resonated preferentially at the alpha rhythm, and propagated waves at a velocity which corresponded to independent estimates for real alpha rhythm in dogs.

The model predicted marked attenuation of wave propagation outside the alpha range for which evidence one way or the other was not available. The choice of the value for the single parameter was arbitrary.

## Freeman

The above early work of the Amsterdam group ran roughly parallel to the early work of Walter Freeman, who was tackling the problem from a very different perspective: that of explaining the EEG characteristics of the olfactory system. He and his colleagues chose to start on the olfactory system because the modeling of its dynamics seemed a comparatively simple task compared to the whole brain. The olfactory bulb is not only much smaller than the whole brain, and relatively isolated from the larger system, but also has a very characteristic EEG. His extraordinary work is difficult to summarize (see his summary in the next chapter, 10b). A particularly important aspect is the way in which his group's physiological observations have guided modeling, and vice-versa.

Olfactory EEG comes in two types: a sharply rhythmic activity at about 40 Hz, during each sniffing inhalation made by the animal, and a tumbling, low-voltage, incoherent pattern apparent between sniffs.

Freeman set out to generate these features and others. His general method of constructing aggregates of inhibitory and excitatory cells was like that of Wilson and Cowan, and van Rotterdam and Lopes Da Silva. Precise physiological measurements were made to provide the relevant transfer functions for excitatory and inhibitory post-synaptic potentials, and the nonlinear relation of dendritic depolarization to output pulse density. These two types of function are termed by Freeman the pulse-to-wave and the wave-to-pulse conversions. The wave-to-pulse conversion is, in his model, a function of the degree of cortical activation. He has also provided a theoretical rationale for the form of the functions.

A very detailed model of the dynamics of the olfactory bulb has emerged — a model which not only accounts for the inhalation and background EEG rhythms, but for the organization of the spatial characteristics of the EEG. The impact of attention and learning upon these patterns has been experimentally demonstrated. Freeman has abstracted from his findings a sophisticated account of the flow and processing of information in the olfactory system, and for the attractor dynamics of the bulb.

A striking general property has emerged in his model, with consequences for all other brain dynamics modeling. That is, the emergence of numerous basins of attraction, each accessible selectively, depending on the particular input stimuli. It appears that the basin of attraction entered also depends upon the state of the system in other respects such as overall excitatory tone, attentional set, and learning. This interpretation fits the physiological data very well, although the full extent of these modulatory influences cannot yet be fully accounted for. Freeman's olfactory model is couched in terms of chaotic processes, and he offers evidence that each olfactory percept may be linked to a particular basin of attraction for the cells in the olfactory system.

Exact demonstration of the differences between attractors has not been fully achieved, most likely because definitive proof of the presence or absence of chaos in a complicated signal is elusive. A part of the difficulty is that the olfactory EEG, if chaotic, must represent not a persisting chaotic process, centered on a single attractor for a prolonged interval, but a transient condition, which only briefly dwells within a single basin of attraction. Freeman has identified basins of attraction with the perception of various odors in relation to their current significance to the animal, rather than the simple stimulus property of the odor. The 'resting' state of the olfactory bulb is within a distinct basin of attraction out of which the state-trajectory is impelled by an incoming odor signal, to one of the many possible other basins of attraction.

Although the experimental demonstration of olfactory chaos has not proved conclusive as yet, the indirect case made is compelling. This work raises immediately the question of whether EEG from the cerebral cortex is chaotic also. Freeman believes that it is. In a recent paper he undertook the transfer of information gained in his olfactory work to prediction of the likely situation in the cortex. In part summary, his views are as follows:

The 40 Hz rhythm of the inhalation olfactory rhythm arises from the cyclic interaction of excitatory and inhibitory cell populations, as a consequence of delays imposed by (or measured as) the pulse-wave conversions. Essentially, the excitatory cells drive the inhibitory cells with about a 5 millisecond lag between the onset to the peak effect. The inhibitory cells, which are partly coupled to the excitatory cells driving them, return inhibition with the same delay. If both systems of cells are excited externally to near their thresholds, an oscillation of excitation and inhibition develops. Exactly the same should hold true of excitatory/inhibitory interactions in the neocortex. This, Freeman believes, explains the correlation of the ECoG with cell firings at 40 Hz in the neocortex. He reasons that the sharp nonlinearities of cell interaction must impose chaotic dynamics in the network in both the bulb and the neocortex. He further points out that the distribution of cell firing rates in neocortex also resembles that expected in chaos, as does the smeared frequency spectrum of the desynchronized EEG.

## Nunez

This worker and his colleagues have developed a series of models of brain dynamics remarkable for their thoroughness of theoretical development from first principles of linear systems theory, again relying upon continuum modeling.

They make similar assumptions regarding the local dynamics of neuronal interaction to those of van Rotterdam and Lopes da Silva. They linearize the output response of a cell mass to a given degree of dendritic depolarization, and make the anatomically well justified assumption that inhibitory interactions are short range, versus the greater range of excitatory/excitatory interactions. The principal extensions then made are:

i) Consideration of the role of long cortico-cortical (excitatory) fibers, which introduce significant delay because of the conduction times of transmission in axons. Indeed, for most of their work they consider dendritic delays as small compared to the maximum axonal delays. One of the effects of this assumption is to make the local field potential response linear with regard to the afferent action potential density.

ii) Consideration of the boundary conditions of the brain. Since the total cortical surface can be considered as a deformed sphere, wave activity of long range will be influenced by the fact that waves can travel around the brain and re-enter their point of origin from opposite directions.

iii) Explicit consideration of the scaling effects of electrode size and situation upon the type of EEG activity recorded.

Development of Nunez's theory has been stepwise, considering first an infinite one-dimensional brain, then a circular brain, a sphere, and a deformed sphere. These different boundary conditions are each considered under differing assumptions of connectivity, input, and settings of various parameters across the ranges permissible in the present (limited) state of certainty of their true values. The development of this work is not yet complete, but the upshot is that for most possibilities within the range of parameters considered, the mathematical models predict the occurrence of standing waves in the brain. The standing waves of cerebral electrical potential are similar in principle to the standing waves generated on a plucked violin string, or a struck drum-head. The predicted form of these standing waves is complicated, and under differing constraints quite different wave shapes are generated, but the following conclusions are emphasized by Nunez:

(a) The Amsterdam model can be incorporated as a local wave description, embedded in the global process, as could any other local model with compatible assumptions as to linearization and time-course of dynamics.

(b) Alpha rhythm for brains of human size is predicted to occur as a global standing wave: the first harmonic of the global system, analogous to the first harmonic of an organ pipe or any other resonant system generating standing waves. This property is additional to any local alpha generation predicted by the Amsterdam group.

(c) The velocity of alpha waves of the global type is predicted to be in the range of six to nine meters per second in the human case. This prediction has been confirmed by methods using both EEG and MEG (magnetoencephalogram).

(d) Halothane rhythm (a cortical rhythm found in anesthesia) is accounted for. This rhythm behaves like a global resonance varying in frequency with the depth of anesthesia, and hence with the velocity of axonal conduction. The frequency of this rhythm is appropriately different for dogs and humans, according to brain size.

(e) The frequencies of the other major cerebral rhythms correspond approximately to those predicted for resonant modes (types of standing wave) of a deformed sphere.

(f) Human alpha rhythm varies with head size in the fashion predicted for corresponding variation in brain size.

(g)  In certain conditions EEG topography indicates the presence of global patterns of activity having large domains of activity all in phase, with a sharp 180 degree reversal of phase between in-phase domains — a feature characteristic of standing waves.

An apparent weakness of Nunez's case is the general similarity of the EEG activity from mammals of widely varying brain size. Nunez deals with this objection by appeal to the presence in the EEG of both local dynamics, and global dynamics: the former in common between species, and the global predominating in large brains. The making of this proviso raises concern as to the conditions in which global resonant modes and local activity are supposed to predominate. At this time no resolution on this problem has been reached, and it has not yet been proven that global resonant modes make a very large contribution to the EEG, if they do so at all. Notably Nunez's mathematical formulations depend on delay times in dendrities which are small compared to delays of transmission in axons as a condition for the emergence of global resonant modes. This condition may not be met in actuality.

## The Australasian group

All of the models described above bear close resemblances in the form of the equations used to construct brain interactions but they differ drastically in the dynamic class of the brain activity they predict. Most notably, Nunez's theories and those of Lopes da Silva *et al.* view the EEG as a linear wave process, and do not address the attractor dynamics of the cortex in any detail. The assumption of linearity is in drastic contrast with the findings of Freeman's group, which suggest strongly that the EEG of the olfactory bulb at least is a chaotic, highly nonlinear type process. This is, at first sight, a serious problem, and might be taken to indicate that the models are completely logically incompatible, rather than partial descriptions of an underlying unified process. Concern for this problem formed the basis for the earliest stages of the work of the author and his colleagues.

We began work with an intention to define the impact upon the EEG of the reticular activating system of the brain. This vital system regulates changes between sleep and waking, partially directs the field of attention of the subject, and produces much more profound changes in the EEG than does any direct sensory stimulus. More important still, interactions between cortex and the activating system form a constant concomitant of consciousness. Activity in the cortex controls the activating system, and the activating system controls the cortex. This striking aspect of brain function receives little detailed consideration in the models above, and this was a defect we set out to remedy — while at the same time attempting to clarify the kind of dynamics which best apply to EEG activity.

We began by making some assumptions about the oscillatory interactions of excitatory cells in the cortex with their inhibitory surround cells, and regarded each pool of such local activity as coupled to that of many others by longer-range couplings, in the same general way applied in the preceding models. We accepted that the interactions of cortical neurons were extremely nonlinear on the microscopic scale, but sufficiently perturbed by noisy inputs from the reticular activating system, that the whole network could be considered as a system of stochastic oscillators. That is, we assumed that local neuronal activity has a large random component rather than being purely chaotic. Deliberately allowing for a high level of random activity in neuronal interactions offers a way of resolving some of the apparent contradictions with which we were faced. We could show that the resulting low-frequency EEG waves obtained from averaging thousands of cells, would tend toward linear wave properties, and toward near-equilibrium dynamics, at large scale. 'Near-equilibrium dynamics' means that the energy of oscillation is approximately equally divided among all the system's modes of resonance.

Wave linearity and near-equilibrium properties are not generally characteristic of physical processes involving high nonlinearity, but it is quite possible for a single system to exhibit different dynamics at different scales. There are numerous other physical examples. For instance, the movement of individual molecules in the air is chaotic, but sound waves traveling through the air are linear waves. This distinction of dynamics by scale (microscopic nonlinearity blurring into global large-scale linearity) does not occur in all systems. For example, workers in Japan have shown that coupled microscopic systems which are individually chaotic, generally preserve chaotic dynamics at the macroscopic scale. A factor (perhaps one of many permissive conditions) permitting different dynamics at different scales appears to be the presence of additional diffuse noise at the microscopic scale.

A series of physiological experiments to check the likely underlying dynamics of the waves led us to believe that large scale EEG activity is mainly linear and near-equilibrium. This gave us some confidence that we could proceed to modeling using linearized continuum models. This in turn implies commitment to population coding of information transfer within the cortex but carries an important proviso — that we could not disregard the interactions which must take place between events occurring at different scales in the brain. Should any small group of cells begin firing rapidly and interacting very strongly with each other, they would locally violate the assumption that interactions between the cells could be considered stochastic. The local, highly nonlinear activity would then emerge as a source of driving signals, governing the generation of waves in the surrounding, stochastic medium. Conversely, the convergence of inputs from millions of cells in the surrounding cortex would be the necessary trigger of this local violation of random interactions. Thus macroscopic and microscopic scales would interact.

These broad guiding principles imply a commitment to a population coded, continuum modeling approach, as opposed to a frequency coded and neural network approach. From this beginning we have subsequently developed a system of simulations which derive from concepts introduced in all the models discussed above. We (Wright and Lilley, 1996; Rennie, Robinson and Wright, 1999) have emphasized certain rapid physiological feedback processes at synaptodendritic junctions which have been under-emphasized in most earlier models. We (Robinson *et al.*, 1999) have recently added consideration of the interaction of cortical and thalamic systems.

Combination of these features considerably increases the match of the simulations to experimentally observed data. At the time of writing our simulations give an account of the theta, alpha, beta, and gamma rhythms, the form of the average EEG power spectrum including its '1/f' background form, 40 Hz oscillations, traveling waves in the cortex, and synchronous oscillation. The model is currently being extended to account for cortical evoked potentials, and for the EEG properties seen in the sleeping state, and we are attempting to apply the same principles to simulations incorporating more detailed anatomical and cellular features.

These simulations thus successfully account for a considerable range of observed physiological properties, and since the essence of the simulations' properties can be expressed in state-equations, we can also use these to describe the phase space associated with the model dynamics and the attractor structure of this phase space. Essentially, the associated phase space is one in which lower levels of cortical activation are associated with point attractor properties, and high levels of activation and some gamma rhythms are associated with point repellors, or limit cycles.

We can now attempt to put a provisional numerical account of brain dynamics together, based upon our current simulations' properties and the associated wave and attractor features which are exhibited.

## AN ACCOUNT OF BRAIN DYNAMICS

Figure 10a.1 shows how brain dynamics can be conceived of as processes embedded within each other at multiple scales.

Firstly, at the microscopic scale in any locale of cortex, local interactions between excitatory and inhibitory cells take place. The mixture of excitatory and inhibitory cells force oscillation upon the local interactions, as the level of overall excitation is increased — much as Freeman explained for the olfactory system. The oscillation is mediated by fast neurotransmitters (see chapter 6, The Brain's Chemistry), and comes about because there is a delay between predominance of local excitation and return inhibition, so that cell firing rates see-saw back and forth. The dynamics of this oscillation may be chaotic activity, with considerable overlying noise. As the level of excitation becomes higher, the frequency of this oscillation becomes higher, and eventually the oscillation reaches a strength at which the cell firing becomes autonomous. That is, while at lower levels of excitation removal of external inputs would cause the activity to quickly stop, at sufficiently high excitation the process begins to run away toward very high firing rates. The local dynamics may now be chaotic or limit cycle, and are less perturbed by background noise. The frequency of oscillation settles around 40 Hz, and the exact attractor state into which the cellular activity enters will be a function of the specific inputs, as in any attractor neural network. The autonomous activity would go on to complete runaway unless some active process brought it back under control. Present indications are that the suppression of oscillation once it has become autonomous is itself a complicated process, involving slow-acting neurotransmitter and interactions between the local cortex and the brainstem.

Secondly, at a scale between millimeters and centimeters of cortex, patches of active cells just below or above the transitional level of excitation which leads to autonomy, participate in interactions which produce synchronous oscillations. The exact mathematical basis for the occurrence of synchrony is not easily grasped, and to add to the difficulty, there would appear to be two somewhat separate bases for synchrony: one nearly linear and operating below the level of transition, and another highly nonlinear, and operating above the level of transition. But the basic mechanism is

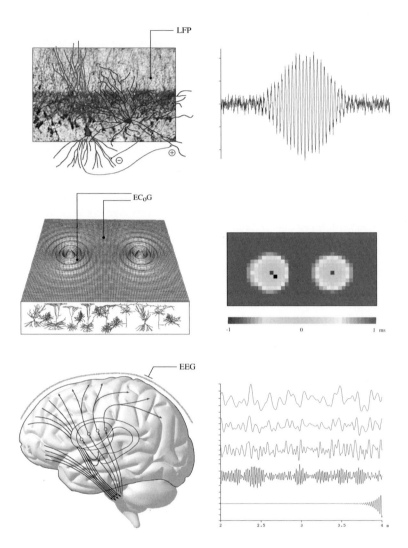

**Figure 10a.1**. The top left panel shows two representative cells within the cerebral cortex — one (an excitatory pyramidal cell) and the other (an inhibitory stellate cell). Populations of these cells are linked together densely in the cerebral cortex. When a small area in the cerebral cortex becomes sufficiently excited by its inputs from elsewhere in the brain, the excitatory and inhibitory cells at a focal site in the cortex enter into cooperative interactions which generate bursts of activity at around 40 cycles per second (40Hz), as shown in the graph of the local field potential (LFP) in the top right panel. This activity is a type of limit cycle.

The middle panels show how at a larger scale these foci of excited cortex generate waves of cortical electrical activity spreading into the less excited surrounding cortical tissue, which radiate outward from the foci, interpenetrating with waves from other sources (middle left panel). The resulting wave activity can be analyzed by cross-correlation, as shown in the middle right panel. Here the lag time for maximally correlated activity (with reference to the recording site shown in the left middle panel) is displayed for the extended field. It is seen that the foci of activity have entered 'synchronous oscillation' (the cells at co-active sites are emitting electrical activity in harmony with each other). It can be shown that information is being shared between the co-active sites.

The lower panels show the overall brain, and EEG activity as generated in simulations from low frequency theta activity, through the alpha, beta and gamma ranges, to 40 Hz. These progressive changes in frequency content reflect the overall level of cortical excitation. At the highest levels of excitation, the self-excited cortical state described in the top panels has been reached.

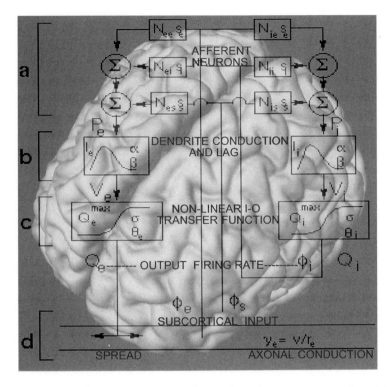

**Figure10a.2.** An outline of the key parameters in the Wright, Robinson, Rennie et al. model. A list and description of many of the model parameters (including their value) serves to illustrate their realistic anatomical and physiological dimensions, and is provided below:

| Parameter | Description | Value |
|---|---|---|
| Nee=Nie | Dendritic synapses from pyramidals | 4120 |
| Nei=Nii | Dendritic synapses from interneurons | 800 |
| Nes=Nis | Dendritic synapses from subcortex | 80 |
| see=sie | EPSP size due to $\phi$e when Ve,i=Ve,i | 2.4 $\mu$V s |
| sei=sii | IPSP size due to $\phi$i when Ve,i=Ve,i | -5.9 $\mu$V s |
| ses=sis | EPSP size due to $\phi$s when Ve,i=Ve,i | 2.4 $\mu$V s |
| Ve,i $^{(o)}$ | Potential at which sep=sip are estimated | -0.060 V |
| Ve,i $^{(rest)}$ | Rest potential ($\phi$e=$\phi$i=$\phi$s=0) | -0.060 V |
| Ve=Vs | Reversal potential for AMPA channels | 0 V |
| Vi | Reversal potential for GABAa channels | -0.070 V |
| $\alpha$ee=$\alpha$es | EPSP decay rate in pyramidals | 68 s$^{-1}$ |
| $\alpha$ei | IPSP decay rate in pyramidals | 47 s$^{-1}$ |
| $\alpha$ie=$\alpha$is | EPSP decay rate in interneurons | 176 s$^{-1}$ |
| $\alpha$ii | IPSP decay rate in interneurons | 82 s$^{-1}$ |
| $\beta$eq | Rise rate of PSPs in pyramidals | 500 s$^{-1}$ |
| $\beta$iq | Rise rate of PSPs in interneurons | 500 s$^{-1}$ |
| $\eta$ | Cut-off frequency for feedback | 200 s$^{-1}$ |
| Qe | Maximal pyramidal firing rate | 100 s$^{-1}$ |
| Qi | Maximal interneuron firing rate | 200 s$^{-1}$ |
| $\sigma$e,i | Sigmoid width parameter | 0.005 V |
| $\theta$e,i | Firing threshold | -0.052 V |
| $\phi$s | Firing rate of subcortical neurons | 10 s$^{-1}$ |
| $\gamma$e | Excitatory spatial damping rate | 80 s$^{-1}$ |
| $\gamma$i | Inhibitory spatial damping rate | 10 s$^{-1}$ |

simple enough. Traveling waves spread out from each patch of cortical activity, generated by the rapid firing of cells at each locale. The waves are stochastic linear waves traveling through the surrounding domains of less active cortex, and it is these waves which are generally observed as EEG, since their wide extent and large scale causes them to predominate in standard recordings. The component of the traveling waves of greatest importance is that mediated by fast, direct, axonal couplings between the two active locales. At each locale the dendritic membranes sum up both the synaptic activations arising locally, and those arriving from the more distant locale. When both sites are active and almost in phase at any particular frequency, this summation is additive. When both sites are out of phase with each other at a particular frequency, cancellation occurs. The net effect is to remove out-of-phase components, so that the two locales fall into synchrony with each other.

The creation of extensive fields of synchronous oscillation locales, each in activity within a particular local basin of attraction, generates a much larger coupled and synchronously oscillating field. The larger field itself will conform to a particular collective basin of attraction and some specific unified concept (grandmother) may be thus represented.

Thirdly, at the scale of the whole brain, the total field of synchrony and its associated basin of attraction generates a pattern of specific signals which descend from cortical to subcortical systems: and the subcortical systems generate signals which return to the cortex. A part of this cortical/subcortical interaction is via the thalamus, and is itself rhythmical. This appears to be the principal way alpha rhythm is generated. But perhaps the more important process is that mediated by descending frontal and limbic mechanisms, and the return influence then exerted by reticular activating system and catecholaminergic neural systems. By this means, the cortex can exert control over the pattern of its own excitation, on the largest scale. Thus, as a function of its current state, the cortex restricts the possible patterns of synchronous oscillation which can come about — favoring some possible attractor states over others. It is tempting to equate this self-restricting process to the direction of attention, and the governance of motivation.

It can be seen that in this view the dynamics of the brain is different in type at each scale: noisy and chaotic at microscopic scale, stochastic and linear for the most part at medium scales. Each area in the cortex is capable of transitions between quiet and passive background states, and highly active patches of briefly autonomous local activity. The whole forms an enormous attractor neural network, with interwoven fields of local activity each conforming to local attractor dynamics, and assembled by synchronous oscillation into coherent groupings of neurons at all scales. The large scale activity controls the microscopic activity, and vice-versa. The totality forms a self-controlling system — all parts control all others to greater or lesser extents.

## BEYOND BRAIN DYNAMICS, TOWARDS SELF-ORGANIZATION AND ADAPTATION IN THE BRAIN

If the account of brain dynamics given above adequately represents the brain's dynamics, are we any closer to an overall explanation of how the brain acts adaptively? Do the types of dynamics exhibited tell us anything about how the brain might organize the storage of information, and its recall? Is any inevitable property implicit in the way systems of this type operate which could enable performance of the massively parallel computations required in all cognitive tasks? Could some innate system of reward and punishment be included, so that the system could learn without detailed supervision?

There are already some fascinating hints that answers to these questions are attainable.

## Chaos and Learning

Most neural network models used in computer science depend upon symmetrical connections between the 'neurons' in the simulation. It is this symmetry which allows the powerful Hopfield and PDP networks to exhibit point attractors, each of which represents a 'memory' or classification of inputs. When connections between neurons are highly asymmetric, most of the attractors in a network are chaotic. However, studded within this sea of divergent flows lie limit-cycle attractors.

Parisi has studied networks of this asymmetric type, and has pointed out that they are well suited to store memories in a way which appears very physiological, since storage can depend upon a phenomenon which has been the subject of much physiological research: that of long-term potentiation (LTP).

LTP describes the way in which the synaptic connections between two neurons become more efficacious when the cells have both been above threshold, and firing, at the same time. In 1949 Hebb suggested that a mechanism of this type could form the basis of learning — a proposal he made before LTP was discovered, and at a very early stage in the development of neural network theory (see the big picture and psychology chapters).

A problem of long standing in implementing Hebbian learning in neural nets has been that the storage of increasing amounts of learning via this type of synaptic modification requires an infinite range of synaptic strength. The ultimate result is also a network which is saturated, and in which basins of attraction begin to fatally overlap. Further, the only physiological candidate to produce Hebbian learning – LTP, is a process which fades away with the passage of time if the cells are not repeatedly coactivated.

Parisi pointed out that asymmetric nets, which have chaotic dynamics in the main but with domains of limit cycles, are well suited to produce a type of memory storage which is accompanied by a process of forgetting. The dual process he termed 'palimpsest' memory.

Palimpsest memory works as follows: sufficiently strong and persisting inputs drive the network into an orbit in state-space within a limit-cycle attractor. The pattern of repeated firing in the limit-cycle strengthens the connections in the cycle, and deepens the basin of attraction. Chaotic activity outside the attractor produces little learning, since repetitive sequences of cell excitation are not present in chaotic trajectories. Unless a limit cycle is repeatedly activated, it too is gradually overwritten with the passage of time. Thus the dynamics and physiology of real neurons appear well suited to providing a palimpsest type of memory.

But how might this process be linked to the brain's overall activity to form a coordinated whole? And what about learning of memories which are not forgotten?

## The 'edge of chaos' and information processing

Models of brain dynamics include dynamic events falling into two different classes (e.g. limit cycles versus stochastic activity) organized so that multiple attractor states involving energy transformations can occur. An example of such is seen in figure 10a.1 (top panels) where at a critical level of excitation local cortical activity bursts into energetic local firing. Such behavior is analogous to phase transitions (e.g. between liquid water and steam) seen in physical systems. Phase transitions may be vital to the operation of the brain, as has been argued from much more general grounds by Langton, to reach the following conclusion:

If a spatially extended system is so organized that portions of the system may go into one phase, while elsewhere the reverse transition is taking place, and the whole is held in a critical balance, these are conditions in which information may be exchanged at high rates. System behavior is characterized by a balance between complete loss of long-term predictability and complete lapse into an invariant, forever predictable, state. This Langton has termed 'the edge of chaos', and his formulation provides an exciting hint as to how the brain may handle information processing, including indications that systems of this type can be capable of universal computation.

## Overall cortical dynamics and control

To follow through the implications of 'edge of chaos' information processing, we have again to think of the large picture of flow of information in the brain, as pictured in the light of the dynamic model described in figure 10a.1.

Sensory information arrives at the special sensory cortices after much preprocessing in lower centers. Signals spread out into the cortical neurons, generating the fields of average activity revealed in the EEG. At certain points of intersection of these waves, the net depolarization, or equivalently, the density of synaptic action, is strong enough to drive local activity to high firing rates, and thus into limit cycles. This has three consequences:

(i)     The local activity 'recognizes' (i.e. classifies) the input by the very fact of having entered a particular limit-cycle attractor, which means local activity has entered a specific basin of attraction. Cellular activity within an attractor basin helps to define the synapses which will be strengthened under the Hebb rule. LTP refreshes the synapses involved, and thus retains a short-term memory, and perhaps helps engrave a new long-term memory via secondary effects on synaptic growth.

(ii)    Local activities become unified via synchronous oscillation to represent 'objects' which arise from both external stimuli and internal states.

(iii)   The high rates of cell-firing involved in the local activity begin to threaten the stability of the local system, and inhibitory local and global mechanisms bring the firing rate down. Thus a phase transition has been negotiated in both directions. When the interaction of many such sites is considered the entire system is on the edge of chaos.

The restabilization of local areas which undergo phase transition must be achieved in part by interaction of the cortex with the brain-stem activation pathways. This balancing act between cortex and subcortical systems may form the substratum to attention, and perhaps mood, both of which are mediated by two-way interactions between the cortex and lower systems.

Within this sketch of the brain's local and global cortical dynamics can be discerned not only the possibility of powerful self-control, but if Langton is correct, the model system falls into a class which can exhibit the capacity for universal computation, and need not be merely stimulus bound. But this is still not enough to account for adaptive behavior in general.

## Self-organization, synchronous oscillation, and coherent infomax

A further concept appears to lead closer still to a schema for brain adaptive learning. As we have discussed, synchronous oscillation is considered by many to be the basis of moment-to-moment association processes in the cortex, but its role may be more basic than that. Kay, Phillips, Singer and others have described a principle applicable to simple neural networks in which the property of zero-lag synchrony between concurrently active sites of input is assumed. Using principles from information theory one can then show what learning rule needs to be applied within the network to maximize the storage of information about *relations* between different input stimuli, rather than information *about* the stimuli. This they term 'coherent infomax'. It turns out that the rule required is a variant of the Hebb rule, with similarities to some of the properties of LTP. Further support for the notion comes from work by Alexander and others, which applies coherent infomax to the establishment via learning of connections within the visual cortex. By this means the connectivity of the real cortex can be closely approximated.

It appears that coherent infomax might be generalized to operate across all classes of activity in the brain, from sensory to motor, and all stages in between, including emotional information, pain, etc. In broad principle, it may be seen that this may lead to the extraction of all relationships between stimuli in the environment, the organism's own responses, and the response consequences for the organism.

## The need for internal motivation mechanisms

It seems then, that our simulated brain belongs among systems which naturally are capable of general computation by virtue of their 'edge of chaos' dynamics. Memories might be stored and accessed short term, and under operation of a local law of synaptic modification there is latent capacity to learn about the relations among all stimuli in the environment, including the consequences of motor activity.

A system capable of utilizing these general properties, and which might approximate learning and adaptation in a natural world, would also require the prewiring, so that certain subsets of all possible stimuli and actions will reflexively lead to approach behaviors and some to withdrawal behaviors. The existence of just such innately wired judgement mechanisms is demonstrated in the operation of reward and punishment pathways which travel from brainstem to cortex via the hypothalamus, and can be recruited by electrical stimulation — as in the famous experiments of James Olds and others. As some natural stimuli led reflexively to these responses, this would tend to produce a partitioning of the brain's state-space into patterns of activity which are 'good for me' and some which are 'bad for me'. Then, thanks to coherent infomax, the development of systems of response to stimuli in themselves neutral, but more distantly associated with either 'good' or 'bad' objects would take place. This would act so as to increase pleasurable approach behaviors, and make escape behaviors maximally efficient — and would yield cognitive maps of all aspects of the environment in survival terms.

Related concepts on neural adaptation have been advanced by Edelman, who calls the principle 'neural Darwinism'. This term emphasizes a deep relationship between the processes of evolution operating upon species, and those operating on the synapses as an animal's private adaptation to its environment evolves.

## FUTURE DIRECTIONS

We have here reached the bounds of rational speculation about the brain's self-organization and its relation to adaptive behavior — and may be yet wide of this goal. However, it should be clear that concepts from physiology, information theory, and the theory of dynamics can be woven together to indicate the general shape of theories to come. It is quite impossible to say whether the full operation of the brain can be comprehended within these terms, even in principle — but this cannot be known until the effort to do so has been made.

As dynamical simulations of the brain improve, they constantly pose demands for more and more accurately specified anatomical and physiological details. In the morass of things we *might*, and do, measure in the brain, theory helps us know what we *should* measure to make better simulations. We may hope that an iterative process has begun, by which theory and experiment can strongly interact, to lead us to increasingly secure and more integrated theories of the brain.

### References and suggested further reading

Amit, D. (1989) *Modeling Brain Function*: The world of attractor neural networks. Cambridge University Press.

Edelman, G. M. (1987) *Neural Darwinism*: The theory of neuronal group selection. Basic Books.

Freeman, W. F.(1975) *Mass Action in the Nervous System*. Academic Press, New York.

Langton, C. G. (1990) Computation at the edge of chaos: phase transitions and emergent computation. *Physica D*, **42**, 12–37.

Nunez, P. (1995) *Neocortical dynamics and Human Brain Rhythms*. Oxford University Press.

Phillips, W. A and Singer, W. (1997) In search of common foundations for cortical computation. *Behavioral and Brain Sciences*, **20**, 657–722.

Rennie, C.J., Robinson P.A., Wright J.J., (1999) Effects of local feedback on dispersion of electrical waves in the cerebral cortex. *Physics Review E*, **59**, 3320-3329.

Robinson, P.A., Rennie C.J., Gordon, E., Bahramali, H., Wright, J.J. (1999) Prediction and analysis of electroencephalographic spectra in terms of physiology (submitted).

Wright, J. J and Liley D. T. J. (1996) Dynamics of the brain at global and microscopic scales: Neural networks and the EEG. *Behavioral and Brain Sciences*, **19**, 285–295.

# BRAIN DYNAMICS:
# BRAIN CHAOS AND INTENTIONALITY

Walter J Freeman

Department of Molecular and Cell Biology, University of California at Berkeley

The search for an understanding of how human beings create themselves through their actions arising in their brains inevitably leads to the study of chaos. Deterministic chaos is characterized by complexity that is self-organized in accordance with simple underlying rules. Examples are found at all levels of organization of nervous systems. The search for simple rules is one good reason for using the tools of the theory of chaos to model neural functions. The appropriate level for researchers who are interested in perception, cognition and consciousness is the level of macroscopic neural populations, because in animals from insects to man, that is where the organization of perception and goal-directed behavior with respect to the external environment occurs. The dynamics of these populations is shaped by learning from the sensory consequences of intentional actions, and it uses chaotic attractors in the cerebral cortex to provide the background 'spontaneous' activity that is required for the creation of novel trials in trial-and-error learning. Humans learn by, in and through chaos, and the global dynamical structure of the neural populations constitutes the self.

## CHAOTIC DYNAMIC BRAIN FUNCTION

The biology of human brains should be approached first and foremost in terms of how they construct intentional behavior, while enabling humans to know what they are doing, and why.

The theory of nonlinear dynamics has greatly expanded our understanding of neural mechanisms by which large-scale patterns of brain activity are self-organized. The new concepts give us fresh insight into the neurodynamics of intentional behavior, how consciousness emerges within the brain, and how brains regulate their own sensory inflow into the cortex (Freeman, 1995).

The classical stimulus–response paradigm in psychology clearly fails to address the most basic properties of biological intelligence and control, which are its autonomy and its creative powers. Chaotic dynamic systems not only destroy information, they also create it. In previous work (Freeman, 1991) a number of aspects of local field potentials from summed dendritic synaptic potentials (electroencephalograms or EEGs) and axonal action potentials (nerve impulses) were described. These studies also show that brains are chaotic systems that do not merely 'filter' and 'process' sensory input. They actively seek sensory stimuli as raw material from which to create perceptual patterns with awareness that replace stimulus-induced activity.

This review sketches the neurodynamics of a prototypical sensory lobe of the brain, the ways in which it is controlled by the forebrain in the mechanisms of attention, the development of the complex system through bifurcations during learning, the formation of classes by generalization, their access by state transitions, and the role of chaos in constructing new attractors in cerebral cortex as the basis for consciousness.

## NEURODYNAMICS IN A PROTOTYPICAL SENSORY SYSTEM (THE OLFACTORY SYSTEM)

The olfactory bulb is a semi-autonomous unit that interacts with other parts of the forebrain by both transmitting information and receiving regulatory feedback. Its endogenous activity persists after it has been surgically isolated from the rest of the brain, which shows that its basic functions are self-organizing. However, its chaotic activity disappears when its parts have been surgically disconnected, showing that its chaos is a global property that is not due to the entrainment of single neurons acting as chaotic generators. The Gamma rhythm (20–80 cycles per second or Hz) of the inhalation olfactory rhythm arises from the cyclic interaction of excitatory and inhibitory cell populations, as a consequence of delays imposed by (or measured as) the pulse-wave conversions. Essentially, the excitatory cells drive the inhibitory cells with about a 5 millisecond lag between the onset, to the peak effect. The inhibitory cells, which are partly coupled to the excitatory cells driving them, return inhibition with the same delay. Transmission between the populations is by action potentials. If the feedback system of cells is excited externally by an electrical stimulus (an impulse), whether it is given to the olfactory system or to the neocortex, the output is a damped oscillation at a frequency in the Gamma range. This explains the correlation of the EEG with cell firings in the Gamma range in the olfactory system and also in the neocortex.

When brain function is observed during time spans of minutes to hours, the olfactory dynamic mechanism appears to be robustly stable over a wide range of amplitudes. It can be destabilized by excessive excitation, when it jumps to an alternative stable state. This jump is called a phase transition, because it is analogous to the

**Figure 10b.1**. A computer-generated pattern simulates a burst of EEG activity in the olfactory bulb, which is triggered by the presence of an odor that an animal has learned to identify. The pattern defines a chaotic attractor, which is shown embedded in four dimensions: the three Euclidean dimensions defining a torus (doughnut shape) and time as shown by the color of the trajectory (Freeman, 1991).

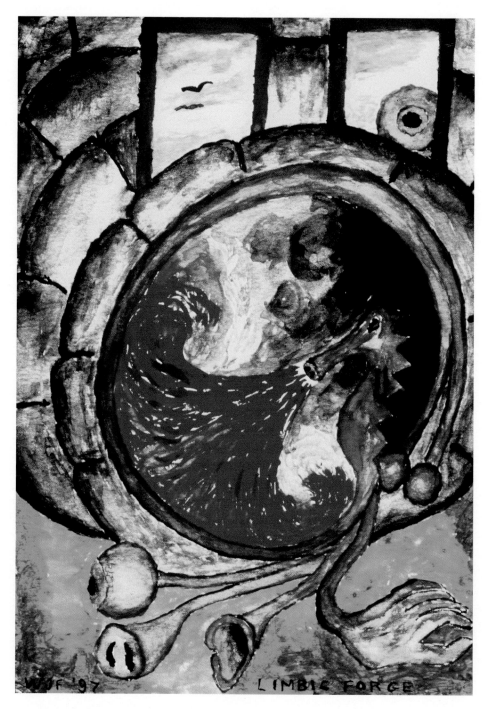

**Figure 10b.2.** The 'Limbic Forge' is a surreal representation of the dynamics of the limbic system, which is centered in the hippocampus ('sea horse'), which sucks up information from the environment through its controls of the eye, ear, nose, and hand. By the dynamics in its action space in the base of the brain, it creates the patterns of understanding that inform the cerebral cortex. These informing patterns provide the windows by which we look into the world, and by which we see ourselves peering into our own windows. *(Painting by Walter Freeman).*

change from a liquid to a solid state, that is, from a relatively disorganized state to a more ordered state. In the olfactory system, the phase transition has the appearance of a change in the EEG from a chaotic, aperiodic fluctuation to a more regular nearly periodic oscillation. In state space the basal background EEG gives a pattern that looks like a bowl of spaghetti. After the transition, the pattern looks like a doughnut. An example of the orderly state is shown in the figure 10b.1, which is a picture of a simulated EEG trace from a model of the olfactory system during an oscillatory discharge during perception of an odor.

The olfactory system can also be destabilized by neurochemical modulatory input from other parts of the brain, as with transition to the stable state of sleep, or with learning. These and related phenomena emphasize the role in neurodynamics of regulatory biases from other parts of the brain onto self-exciting populations in neural system control. Most importantly, learning takes place in rapidly repeated small steps, which cumulatively give the appearance a trajectory in state space that never repeats itself.

## ADAPTIVE DYNAMICS

Some scientists view perception as a late stage of a process that begins with sensory transduction to form representations of stimuli (commonly in the firings of feature detector neurons), proceeds through 'binding' of the parallel activity of multiple features to represent objects, and then by the serial processes of filtering, normalizing, matching with representations retrieved from storage for pattern completion and classification. Perception is completed upon the binding of the representations of an object from the multiple sensory systems, and after an appropriate value or meaning has been attached to the fused image.

An alternative view, in which a percept is a goal-directed action that is organized by large scale neural interactions in the limbic system. Such action is intentional, in that it forms within a framework of space and time that has been constructed from recent and remote experiences of action and its sequella, and it constitutes 'stretching forth' into the world, in order to shape the self in accordance with what is there (Freeman 1995).

The motor commands that issue through the septum and amygdala are accompanied by preafferent 'corollary discharges' sent by the limbic system to all of the sensory cortices, which constitute attention by shaping the dynamic sensitivities of the cortices, in respect to the anticipated changes in sensory inflow that will follow the intended actions.

Thus the sensory systems are already primed to respond in selective ways to the stimuli that are being sought through listening, looking, sniffing, etc.

Closure of the action–perception cycle takes place following the de-stabilization of the sensory cortices, their construction by nonlinear dynamic interactions of spatial patterns of activity, the convergence of these patterns into the limbic system, whence issued the request for input, and finally the updating of the limbic activity. Brain theory suggests that the cyclical process of emergent goal-seeking, preference, and sensory feedback constitutes the basis for what we perceive subjectively as attentiveness leading to consciousness. This inference is supported by the state of 'absence', when the normal chaotic dynamics is suppressed, for example in an epileptic state. The construction of each pattern is guided by a chaotic attractor, which was formed during learning. Perceptions are triggered by stimuli, but they are shaped by connectivity patterns that were laid down during the past learning, and by neural messages from the limbic system that modulate the attractor landscapes of the sensory cortices. The images of memory are not stored and retrieved as in computer systems. The construction of a pattern following a stimulus is not a representation of the stimulus. It is a retrieval of the history of past experience and present significance of the stimulus for the subject receiving it, which is done by a creative construction in the limbic forge (figure 10b.2). The dynamics revealed in these systems may assist us to understand the mechanisms of perceptual disorders that are found in attentional deficits, hallucinations, and more generally the failure of communication among humans.

## LEARNING AND BRAIN DYNAMICS

Learning in animals usually occurs when a reward or punishment accompanies a novel stimulus. If reinforcement is not given, then habituation takes place, by which the neural system decreases its sensitivity to the stimulus on repeated presentations. The decay rate of the impulse response increases, showing an increase in stabilization. This process is fully reversible. It is likely that learning is always a combination of processes for association and habituation. Input that is wanted is reinforced, and input that is unwanted is habituated. Both processes are necessary for learning. Connection strengths change only in the excitatory net in accordance with a modified version of the Hebb rule. The system does not learn single events but assigns inputs into classes (Yao and Freeman, 1990), corresponding to its generalization gradient.

The output of the bulb is a spatial pattern of amplitude modulation of the common waveform of the oscillation, which expresses the cooperative interactions of all of the bulbar neurons. Correspondingly, each local area in the targets sums the activity that it receives. By virtue of this spatial integration the only activity that is enhanced is that which has the same frequency everywhere in the bulb. In this way the bulb transmits its own created pattern and not the imprint on its activity of the sensory input. The same modes of operation have been found in the visual, auditory, and somatosensory cortices (Barrie, Freeman and Lenhart, 1996).

The conclusion is that the brain can only know what has been constructed by its sensory cortices under its control and guidance. Learning therefore takes place by a structural change in the system, such that its behavior in the future is dependent on past experience.

## A ROLE FOR CHAOS IN HEBBIAN LEARNING

If the inhaled air does not contain a known odorant but a novel chemical, then bursts fail to form, and the spatiotemporal patterns are disordered. Activity does not conform to any pre-existing spatial pattern. This suggests an important role for chaos in Hebbian learning. Strengthening of a Hebbian synapse requires concomitant activity in both the presynaptic and postsynaptic neurons.

A chaotic generator appears to be an optimal way that the bulb can provide novel spatial patterns of neural activity, which can then be consolidated by the learning process. Much remains to be learned about the neural dynamics of perception and cognition, and their emergence as the outcome of intentional actions created by the limbic system. These studies already show that the way to comprehend complex systems is to understand the rules by which they organize themselves, using their own endogenous chaotic activity for the creation of novel patterns.

## DIRECTIONS FOR FUTURE RESEARCH

Enjoyment of the full benefits from nonlinear brain dynamics will require extensive development of the basic neuroscience. There are three areas in particular need of attention. The first area is study of the interrelations between the microscopic activity of neurons, as measured by their action potentials, and the macroscopic activity of populations, as measured by their dendritic potentials. An example is provided by work on the auditory cortex of the Mongolian gerbil, for which the neuroanatomy and the behavioral and tonotopic properties of unit responses to FM (frequency modulated) tones have been well established (Ohl and Scheich, 1996), and for which the tonotopic properties of the EEG have now been mapped (Ohl et al., 1998). If animal subjects are trained to discriminate FM tones and to generalize the modulations to different parts of the spectrum, then the basis can be laid for direct comparisons of unit activities (now called activities of 'feature detectors') at the microscopic (sensory) level with EEGs at the macroscopic mean field (perceptual) level. This kind of direct comparison, which is requisite for understanding the transfer functions, cannot be made in the olfactory system, owing to lack of adequate control of the odorant inputs on successive trials.

The second area is the study of the interrelations among the primary sensory cortices (PSCs) and the limbic areas, particularly the entorhinal cortex and hippocampus (Smart et al., 1997; Kay and Freeman, 1998). Two general approaches are currently proposed for the integration of multiple sensory inputs in different modalities into Gestalts. The conventional view is that inputs to the PSCs are sent by serial synaptic transmission to a succession of cortical areas, each extracting higher order features, often with relays through the thalamus, until the highest level of analysis is reached in the frontal lobes, where the Gestalt is synthesized. The alternative view proposed here is that all the PSCs are in continuous cooperation with the limbic system, and that the formation of a Gestalt is by successive first order phase transitions involving the entire hemisphere. The evaluation of these two conceptions will be made possible by simultaneous EEG recording from the PSCs and the limbic system during performance of goal-directed activities by trained animals with implanted electrode arrays.

The third approach concerns the dependence of chaotic neural activity on the noise that is contributed by masses of neurons. The relationship of noise to chaos is another example of the circular causality of microscopic and macroscopic activity, because the myriad action potentials of cortical neurons are sustained by global interactions of the neurons that are manifested in the EEGs. The macroscopic state both feeds on and controls the microscopic activity (Freeman, 1996).

The development of adequate brain theory will depend on advances in the mathematics of chaotic dynamics. At present most of the theory has been devoted to the description of low-dimensional, autonomous, noise free systems, such as the Lorenz, Chua and Rössler attractors, which comprise twist-flip maps. Brain chaos is noisy, nonstationary, and non-autonomous. A much better model is provided by the laser, which has been described in the context of synergetics (Haken, 1983). A further development will be required in the direction of analog devices.

Simulations of olfactory chaotic activity patterns have recently made it clear that there is a basic limitation in the use of digital computers to simulate chaotic processes. Chaos has the property of infinite sensitivity to initial conditions, so that chaotic processes are continuous in time and space. Digital computers use binary integers, which digitizes and compartmentalizes representations of continuous events. As the result, chaotic simulations degenerate into quasi-periodic and point attractors, from which they cannot recover. Although the simulations can be improved with additive noise (Freeman *et al.*, 1997), the computational theory is weak, and the way is opened to solving the differential equations by analog simulation (Eisenberg *et al.*, 1989), seeing that noise is essential as well as unavoidable.

## REFERENCES AND SUGGESTED FURTHER READING

Barrie, J. M., Freeman, W. J., Lenhart, M. (1996) Modulation by discriminative training of spatial patterns of gamma EEG amplitude and phase in neocortex of rabbits. *Journal of Neurophysiology,* **76**, 520–539.

Eisenberg, J., Freeman W. J., Burke, B. (1989) Hardware architecture of a neural network model simulating pattern recognition by the olfactory bulb. *Neural Networks,* **2**, 315–325.

Freeman, W. J. (1991) The physiology of perception. *Scientific American,* **264**, 78–85.

Freeman, W. J. (1995) *Societies of Brains.* A study in the neuroscience of love and hate. Lawrence-Erlbaum Associates, New Jersey.

Freeman,W. J. (1996) Random activity at the microscopic neural level in cortex ('noise') sustains and is regulated by low-dimensional dynamics of macroscopic cortical activity ('chaos'). *International Journal of Neural Systems,* **7**, 473–480.

Freeman, W. J., Chang H-J, Burke B. C, Rose P. A, Badler J (1997) Taming chaos: Stabilization of aperiodic attractors by noise. *IEEE Transactions on Circuits and Systems,* **44**, 989–996.

Haken, H. (1983) *Synergetics: An Introduction.* Berlin: Springer.

Kay, L. M., Freeman. W. J. (1998) Bidirectional processing in the olfactory-limbic axis during olfactory behavior. *Behavioral Neuroscience,* **112**, 541–553.

Ohl F. W., Scheich H. (1996) Differential frequency conditioning enhances spectral contrast sensitivity of units in auditory cortex (Field AI) of the alert Mongolian gerbil. *European Journal of Neuroscience,* **8,** 1001–1017.

Ohl, F. W, Scheich H, and Freeman, W. J. (1998) Topography of averaged auditory evoked potentials (AEPs) in gerbil auditory cortex. *Procedings at the 26th Göttingen Neurobiology* Conference. II, 332.

Yao, Y, Freeman, W. J. (1990) Model of biological pattern recognition with spatially chaotic dynamics. *Neural Networks,* **3,** 153–170.

# Chapter 11

# MODELS OF THE BRAIN IN NEUROLOGY

## Victor Fung and John Morris

Department of Neurology, Westmead Hospital and The University of Sydney

Neurology is the branch of medicine which deals with diseases of the nervous system. Behavioral neurology refers to the study of how diseases of the brain affect behavior. In this chapter, we will discuss how careful observations on the effects of diseases of the brain give valuable insights into how the nervous system is organized, thereby contributing to many models across disciplines.

The fundamental principle which guides neurologists in their every day practice is the attempt to relate symptoms and signs to the anatomical location of the underlying lesion (structural change) in the nervous system. This localizationist model has proved most useful in dealing with disorders of the peripheral nervous system, spinal cord, brainstem and primary motor and sensory cortex, where there is a good correlation between structure and function. However, when dealing with other parts of the brain such as the frontal lobes and basal ganglia, the relationship between structure and function is less clear-cut. Here, a marked abnormality in neurological function may not be associated with any obvious structural abnormality of the brain. Conversely, a demonstrable lesion may not be associated with any clinical effects. Similar symptoms and signs may result from lesions in different parts of the brain.

The task of attempting to relate symptoms and signs to a lesion has been made much easier for the neurologist in the last few decades with the advent of sophisticated and safe brain imaging techniques. Structural and functional imaging techniques are revolutionizing our concepts of how the brain works and have transformed the practice of neurology (see chapter 15, Human Brain Imaging Technologies).

Models of the brain derived from clinical practice largely concern themselves with the function of groups of neurons: the interactions between individual neurons and the cellular signals used to code those interactions are still the domain of microscopic scale neuroscientists. The two approaches are complementary. The findings of basic scientists have important implications for the clinician. Conversely, the observations regarding brain function made by neurologists provide a framework particularly to research into the brain as a system.

## Anatomical and developmental considerations

The anatomy chapter outlined the basic architectural model of the brain as comprising sensory input and motor (including basal ganglia and cerebellum) output networks, with a considerable amount of association cortex (linked to limbic system) in between. Details concerning the input and output networks are also provided in the sensory-motor chapter.

The brain thus appears to be organized so that certain functions, sensory and motor, are separately processed, albeit with extensive interconnections, in discrete areas.

Given the diversity of genetic and environmental influences to which any individual is exposed, neurological development in the infant progresses along a remarkably predictable path. This suggests that significant aspects of early development may be pre-programmed (more dependent upon genetic than environmental influences). When examining the normal infant, a number of responses can be elicited. Neurologists commonly refer to these as primitive reflexes; the grasp reflex and automatic walking are examples.

The grasp reflex is characterized by grip formation in response to stimulation of the palm. It is present in the foetus after only 10 weeks' gestation and persists for a few months after birth. It re-emerges in adults with lesions of the frontal lobes. By the age of 6 months, all parents will know that infants spontaneously reach out and grab hold of any interesting or novel object; next it goes straight into the mouth. Groping behavior can also be seen in adults with frontal lobe damage. Automatic walking is seen in newborn infants and persists until up to 8 weeks after birth. It can be elicited by holding the infant upright, leaning slightly forwards, and allowing one foot to make firm contact with the underlying surface.

It thus appears that certain behaviors are pre-programmed and do not need to be learned. They disappear as the brain matures but they are not all lost, for they may reappear if the part of the brain which is suppressing them becomes damaged. The concept of behavior being genetically predetermined should not be too foreign. We are all comfortable with the knowledge that birds learn to fly, or build nests, without ever being taught. Even more vivid is the image of a newborn marsupial, millimeters in length, climbing a distance of many centimeters to reach its mother's teat. The parallels with grasping and groping are evident.

The presence of primitive reflexes suggests that the developing brain is not a pluripotential collection of interconnected processing units. Although the reasons for the existence of these responses is speculative (for example, as survival advantage or remnants of phylogenetically ancient reflexes), they undoubtedly influence the formation of new sensorimotor associations. In so doing, they may help influence developing cortical connections. This concept has been termed *selective stabilization* by Changeux and Danchin and *functional validation* by Jacobson.

PET studies have identified distinct, uniform patterns in the development of human brain during infancy and childhood. Measures of local cerebral metabolic rate for glucose show sensorimotor cortex, thalamus, brainstem and cerebellar vermis to be active in the first 5 weeks of life. By the age of 3 months, parietal, temporal, primary visual cortex, basal ganglia and cerebellar cortex have increased activity significantly. Prefrontal cortex and visual association cortex do not become active until the age of 8 months. The evolution of metabolic activity in the infant is consistent with the clinical and anatomical evidence linking the development of many reflexes or behavior to maturation of particular areas of the brain. Interestingly, overall brain levels of sugar metabolism continue to rise to 2–3 times adult levels until about 9 years of age, at which time they gradually subside to adult levels. The high levels seen in childhood are in keeping with the overgrowth of neurons, synapses and dendritic spines occurring as part of normal brain development.

From these clinical and investigational studies of neurological development, it emerges that early sensorimotor experiences are governed not only by environmental influences, but also by reflex responses which are hard-wired and similar for all individuals. The ensuing sensorimotor interactions may in turn influence brain development. Therefore ontogenetic, as well as anatomical considerations, must be taken into account when modeling brain function.

## HISTORY OF BEHAVIORAL NEUROLOGY

Attempts to localize mental processes to the brain can be dated to antiquity. In the Hippocratic writings from the fifth century B.C., one can find written:

> And men should know that from nothing else but from the brain came joys, delights, laughters and jests, and sorrows, griefs, despondency and lamentations. And by this, in an especial manner, we acquire wisdom and knowledge, and see and hear and know what are foul, and what are fair, what sweet and what unsavory … (cited in Plum and Posner, 1982).

Herophilus of Alexandria, living in the third century BC, dissected the human brain, and concluded that the soul must reside in the ventricles. The physician Galen, in the second century AD, agreed with the beliefs of Hippocrates and Herophilus, but disagreed with Aristotle, who claimed the heart as the organ of sensation.

The ancient Greeks believed *pneuma* (air) to be the vital principle of living things. In the *rete mirabile* (a network of blood vessels present in the brain of some animals, but in fact absent in man) this was converted to *animal spirit*, which was then refined in the ventricles before being circulated as the basis of nervous activity. However, Galen favored the substance of the brain rather than the ventricles as the location of the soul. Theories of brain function underwent little change over the next fourteen hundred years. Although the anatomist Vesalius, in the 16th century A.D., still believed that 'the ventricles bring the animal spirit into being … ' he also wrote ' … but I hold that this explains nothing about the faculties of the Reigning Soul'. However, he refrained from providing an alternative theory, stating ' … how the brain performs its function in imagination, in reasoning, in thinking, and in memory … I can form no opinion whatever' (cited in Spillane, 1981).

It was the English physician Thomas Willis in the 17th century who finally dispensed with the theory of the ventricles. He firmly believed in the importance of the brain's substance, noting the division of the brain into two hemispheres, as well as the structures forming the connections between them. Willis recognized a separation between voluntary motion, which he felt was determined in the corpus striatum, and involuntary motion, which he postulated was cerebellar in origin.

At least four categories of whole brain models are now evident (outlined in A-D).

## A) Localizationist model

Localization of discrete cognitive functions to distinct regions of the brain was first attempted by the French physician Franz Joseph Gall in the last half of the eighteenth century. Gall studied and categorized human behavior, and attempted to correlate psychological function with the size of the brain's various convolutions. His extension of this theory, that the shape of the skull (molded by the underlying brain) could predict behavior and intellect, became known as phrenology. Gall attributed the inspiration for his theories to a recurrent observation made during his childhood, when:

> I continued to notice that those who learn most easily by heart had prominent eyes … I was seized by the idea that eyes thus formed were the mark of an excellent memory (cited in Spillage, 1981).

The widespread dissemination and extension of the theory of phrenology ultimately led to the testing of these notions and Gall's teachings becoming discredited.

Following the work of Gall, many others began to speculate on the localization of cognitive functions. Of particular interest was where the 'center' for speech might lie. In 1861, the French surgeon Paul Broca published a seminal paper describing two patients with impairment of speech but retained comprehension: at autopsy, both patients had lesions of the left frontal lobe. Following similar observations on another eight patients, he concluded that the left hemisphere was responsible for processing language.

Neurologists now began to view the brain from a strict localizationist model perspective: if a lesion could result in a specific defect in cognitive function, it was assumed that that function was processed in its entirety in that region of the brain.

One of the founders of English neurology, Hughlings Jackson, soon pointed out a limitation of that approach. Some patients with left anterior frontal lobe lesions lost their capacity for propositional (meaningful) speech yet produced 'recurring utterances'. For example, the only conversation one of Jackson's patients was capable of was 'Yabby' (Jackson, 1880).

This was repeated again and again. He wrote:

> To say that the disease 'caused' these utterances, a positive condition, is absurd, for the disease is destruction of nervous arrangements, and that could not cause a man to do something; it has enough to answer for in leaving him unable to speak. The utterances are effected during activity of nervous arrangements which have escaped injury (Jackson, 1880).

His contribution was to note that not only do the negative features associated with a brain lesion inform us about brain function, but also the positive features. In other words, Jackson recognized that not only is the ability which is lost an important clue to the function of brain regions, but also the capacities that are retained. Jackson proffered two explanations for the existence of positive phenomena. To account for abnormal positive features, he felt that the nervous system must have a heirarchical arrangement, with 'gradations from the most voluntary to the most automatic'.

To account for normal positive features, such as the retained ability to understand speech in patients with anterior lesions, he recognized the necessity to separate complex cognitive functions such as speech into component parts (for example the perception of images associated with words, the words themselves and propositional speech). Although his hypothesis that comprehension must therefore be processed bilaterally (and in fact predominantly in the right hemisphere) was incorrect, his realization of the limitations of lesion analysis was an important advance. Jackson was one of the first to recognize that complex higher functions cannot be localized to a single area of the brain: the processing of such functions must be distributed.

## B) Holistic model

In the first half of the twentieth century, neurologists were joined and perhaps overtaken in their quest to unravel the mysteries of the human brain by scholars in the related field of psychology. Two figures warrant particular mention. The great Russian psychologist Luria continued the tradition of investigating the workings of the brain utilizing lesion analysis, both in humans and primates. However, he coupled the established techniques with new, formalized techniques of quantitating mental processes. Arising chiefly from Luria's prolific work, the new discipline of neuropsychology was born. Although not alone, Luria was certainly one of the more influential writers who recognized the importance of Hughlings Jackson's views on the limitations and interpretation of lesion analysis. Luria related the hierarchical organization of the brain, a physiological phenomenon, to knowledge about how different regions of the brain are connected (see Luria's model in The Big Picture). He realized that Flechsig's observations, which revealed how neurons in primary sensory, unimodal and multimodal association cortex are interconnected, had profound implications on how information in these areas might be processed. So too did Luria recognize the necessity to regard mental processes in the brain as being distributed and interactive, rather than strictly localized. He wrote:

> ... any human mental activity is a complex system effected through a combination of concertedly working brain structures, each of which makes its own contribution to the functional system as a whole. This means, in practice, that the functional system as a whole can be disturbed by a lesion of a very large number of zones, and also that it can be disturbed differently in lesions in different localizations (Luria, 1973).

In the early twentieth century, a school of psychologists, chief amongst them the American Karl Lashley, championed a more holistic view of brain function. Their motivation was experimental evidence which seemed to contradict the predictions of strict localizationist theories. Lashley was part of an intellectual school which included Flourens, Goltz and Franz. The approach which each of these researchers used was to perform lesion experiments on animals and observe the resultant effects on their behavior. Lashley's experiments on rats, in which he found strikingly little evidence for specific defects of behavior arising from lesions in different regions of the brain, challenged the concept of localization of cognitive functions held by the neurologists of that time. For example, the ability of a rat to negotiate a maze was equally affected by any lesion of approximately the same extent, regardless of where that lesion was made. Lashley concluded that it was the amount of brain damaged, rather than the site of damage, which determined the neurological deficits or behavioral change:

> The term 'equipotentiality' I have used to designate the apparent capacity of any intact part of a functional area to carry out, with or without reduction in efficiency, the functions which are lost by destruction of the whole …

This capacity varies from one area to another and with the character of the areas involved. It probably holds only for the association areas and for functions more complex than simple sensitivity or motor co-ordination.

The equipotentiality is not absolute, but is subject to a law of mass action whereby the efficiency of performance of an entire complex function may be reduced in proportion to the extent of brain injury within an area whose parts are not more specialized for one component of the function than for another (Lashley, 1963).

This view of an equipotential association cortex and a focus on whole brain function, challenged localizationism as the dominant model of brain function.

## C) Disconnection model

In 1965, the Harvard neurologist Norman Geschwind published an article entitled 'Disconnexion syndromes in animals and man' (Geschwind, 1965). His careful review of previously reported cases in the literature, combined with original observations on personal cases with callosal lesions, re-established the concept of the disconnection syndrome, a cluster of symptoms and signs produced by disruption of pathways connecting one area of the brain with another. Geschwind demonstrated that complex cognitive functions *can* be analyzed and broken down into component parts, vulnerable to selective disruption by lesions occurring in small, localized and anatomically predictable regions of the brain. Geschwind's theories affirmed the importance of localization of brain function, but also emphasized the importance of the connections between them. He thus brought together elements of both localization and holistic models.

Geschwind not only made sense of observations from the past, but also preempted future models of brain function. In discussing the apparent lack of expected loss of function in some patients with localized brain lesions, he postulated 'we are dealing with an equipotential system in which a part can take over some of the functions of the whole'. In debating whether there is a single process of recognition he speculated 'there are multiple parallel processes of appropriate response to a stimulus', suggesting that recognition is a summation of sensory response and association, rather than an isolated process. He recognized the possibility that learned motor acts are 'redundantly represented' and might be spared in partial lesions. Thus he predicted many of the facets of the model of brain function which has become known as Parallel Distributed Processing.

## D) Parallel Distributed model

In 1986, two neuropsychologists, James McClelland and David Rumelhart, co-edited two volumes entitled *Parallel Distributed Processing* (Rumelhart *et al.*, 1986). Borrowing from computational strategies and modeling (see chapter on Computer Models), they postulated that the brain uses a similar system of processing. The key elements of a parallel distributed processing system are simple units, with preprogrammed functions and constraints, which are interconnected and interdependent. Such neural networks, although modeled with much slower processing capacities than conventional computers, are much better at dealing with the sorts of tasks confronted by biological systems. This efficiency is gained from the use of parallel and iterative processing loops. In other words, multiple aspects of processing occur simultaneously, rather than serially, with feedback occurring at multiple levels, rather than at the endpoint. Such a system has the advantage of becoming more efficient, rather than less efficient, with a processing task that involves multiple constraints. This is a feature of biological, but not conventional computational systems.

It is conceived that collections of neurons within a small area form local networks which possess or develop functional specificity. Local networks of neurons are overlapping and possess an element of redundancy. Together they form parts of larger systems (multifocal networks). Brain function is the result of summated activity in these networks. Laboratory models of such systems have been shown to have the capacity to reproduce an important facet of brain function: learning, or in other words facilitation of a particular pattern of neuronal activation, through repeated use.

Marsel Mesulam (1998) has reviewed the evidence for such systems in attention, language and memory. He points out that many of the known characteristics of a process such as directed attention lend themselves to interpretation in the light of a neural network model which employs parallel distributed processing. For example, isolated lesions in three anatomically distinct domains of the brain, or the connections between them, have consistently been associated with a deficit in directed attention (for example a neglect of extrapersonal space). Each domain is associated with impairment of different but interrelated behavior, suggesting overlapping local networks which function in parallel. Anatomical studies have demonstrated widespread reciprocal connections between each of these areas, providing a substrate for iterative processing. Both shared and also unique connections between these three regions and other areas of the brain emphasize the distributed nature of this system. However, a significant gap remains between conventional network and parallel distributed models. As Mesulam notes, 'It is more difficult to specify the neurobiological features and computational algorithms for large-scale networks'. Mapping possible neural networks and elucidating the ways in which they function is the challenge which confronts the modern neuroscientist.

## How lesion analysis gives insights into brain function

It becomes apparent from reading about the history of behavioral neurology that lesion analysis has played a vital role in the development of neurological models of the brain. What are the strengths and limitations of these insights?

## The Limitations of Lesion Analysis

The effect of a destructive lesion is dependent upon several factors. This accounts for some of the limitations of lesion analysis.

The stage of brain development at which a destructive lesion occurs will influence its effect. For example, a cerebral infarct late in life arising from blockage of the left middle cerebral artery can destroy language function completely, with little or no recovery. However, following removal of the left hemisphere in right handed infants, language capabilities can develop almost normally. The ability of other parts of the brain to take over a function following damage to that part which, during normal development, would be essential for its processing, suggests a degree of neuronal plasticity. This is an important, but poorly understood, quality of the nervous system.

Another variable is the rate at which a lesion develops. A slowly destructive lesion will lead to brain dysfunction which evolves in character. The pattern of this evolution will depend on such factors as the direction from which the lesion is extending, whether the process infiltrates or merely compresses the underlying brain, and whether the rate of destruction allows any compensatory processes to occur.

No two lesions are likely to be identical, because of variations in the microscopic extent of lesions and the individuals in which they occur.

## The Interpretation of Loss of Function

The most obvious sign of a lesion in the brain is loss of function. If a particular function is consistently impaired by lesions in a particular location, it is reasonable to conclude that that area of the brain must take part in the processing of the missing function. However, just because a part of the brain takes part in processing a certain function does not mean that processing is confined to that region. For example, visual loss identical to that experienced with a lesion of the occipital lobe can result from a lesion of the thalamus. It is correct to conclude that visual processing involves the thalamus, but it would be incorrect to assume that the thalamus was the center for vision.

## The Interpretation of Positive Phenomena

Destructive lesions do not only lead to loss of function or negative symptoms, but also to positive symptoms. Positive symptoms can occur due to release of function, because a lesion has destroyed areas of brain which inhibit or participate in the control of that function. It follows that a balance between activity in different areas of the brain is required for normal brain function. Alternatively, a lesion might serve as an irritative focus, which triggers abnormal activity in neurons near the lesion. In many circumstances it remains unknown which of these mechanisms is operating.

An example of release of function is intermanual conflict, which can be seen (usually transiently) after lesions of the corpus callosum. In this curious phenomenon, one hand can act at completely crossed purposes to the other. For example, such patients might button up his shirt with one hand, only to have his other hand immediately undo the same buttons.

## EXPERIMENTS OF NATURE — A JOURNEY THROUGH THE BRAIN

Diseases which affect brain function can be viewed as experiments of nature. An overall picture of how the brain works can be pieced together by observing what happens when parts of it go wrong. Lets now explore some of the effects of lesions in different parts of the brain.

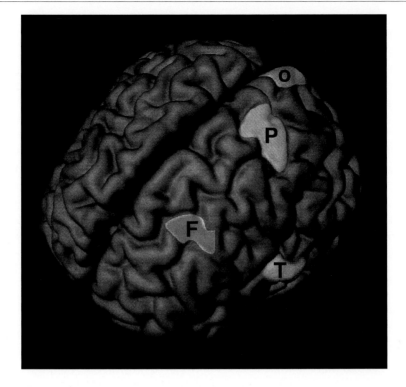

**Figure 11.1.** Lesions in specific parts of the brain (O: occipital; P: parietal; T: temporal; F: frontal lobe lesion).

## The Occipital Lobes

Patients with lesions of the occipital lobes (O in figure 11.1) experience difficulties with vision. Consider what happened to a patient who suffered a stroke which destroyed most of the occipital lobe on the left:

> A 65 year old man suddenly noticed that when he looked at his wife, he could only see half of her face, that half which lay to his left. He had not lost vision in the right eye. *Shut your right eye and look at someone's face; you will see all of it.* The reason that he could only see half the face was that he had lost the right half of the visual field of both eyes.

This patient demonstrates an important principle about how vision is organized — visual information from both eyes goes to each occipital lobe (left and right). Information in the left half of the visual field of each eye is processed in the right occipital lobe, and vice versa. The brain then synthesizes the different information going to each lobe into a single image. This feature of vision — that we combine the two slightly different images from each eye into a single image — allows the development of stereoscopic, binocular vision.

We now know that each occipital lobe receives information from both eyes, and therefore a lesion on one side does not make us blind in one eye. What happens if someone has a lesion of both occipital lobes (a bilateral lesion)? Such patients become 'cortically blind'. Their eyes still work, but the information has nowhere to go to be processed — the result is the same as if they had lost vision from both eyes. A bizarre manifestation of this sort of bilateral lesion is that sometimes the patient's concept of vision is also destroyed. Such a patient may actually deny he is blind. When asked what he can see, he will make things up, even though he cannot see anything at all.

## The Parietal Lobes

Patients with lesions of the parietal lobe (P in figure 11.1) develop three broad categories of problems: loss of feeling (somatosensory perception); defects in attention and spatial awareness; and defects in specialized cognitive functions (such as the ability to manipulate mathematical or spatial concepts). In the case of the first two, the problem only affects the side of the body opposite (or contralateral to) the lesion. The type of feeling that is lost tends to be discriminative sensation. The ability to feel pain, or hot or cold, may be still intact. But such patients lose their ability to tell how far apart are two objects touching their skin, or what they are holding in their hand if they are unable to look at it. The following case illustrates what happens if we develop a more extensive lesion in the parietal lobe:

> A 50 year old woman became unable to dress herself; she could not find the right holes to put her arms in. She sometimes got lost in the shopping center that she had frequented for years. One night she was shocked to find a strange arm in bed with her; she put the light on and discovered it was her own left arm. She only ate food from the right side of her plate and her family had learned to turn the plate round to get her to finish the meal.

When one of our parietal lobes does not function properly, we not only fail to feel things on the opposite side, but also cease to be aware of our body on that side. This is called neglect. Our spatial sense becomes faulty and we get lost easily. The parietal lobes appear to be important in monitoring the objects and space in the world around us. The type of problem which this patient had is most marked in lesions of the non-dominant (usually the right) parietal lobe.

## The Temporal Lobes

The temporal lobes are important for processing hearing, memory and smell. The following cases demonstrate what happens when areas are injured or destroyed (T in figure 11.1):

> A 58 year old man complained that he could not hear; this had developed quite suddenly. He had suffered a mild stroke causing weakness of his left side some 2 years before, from which he had made a good recovery. Now, he found he was unable to understand a simple spoken command such as 'Touch your nose', yet when the instruction was written, he had no trouble understanding it. When played Beethhoven's 5th Symphony, he said he could hear something, but it sounded like a 'loud noise'. He could hear his watch ticking and had no trouble hearing the different pitches of sound used in hearing tests. A CT scan showed large infarcts (strokes) in both temporal lobes.

The patient was not deaf in the usual sense, for he could hear simple sounds like his watch ticking. His problem was that he could not process complex or meaningful sounds such as music or speech. Sounds which we hear are processed in both temporal lobes. In retrospect, one temporal lobe had been damaged by his first stroke. This caused no problem with his hearing because he still had the other temporal lobe. It was only when the remaining temporal lobe was also damaged that he complained of deafness. Put another way, we only need half a brain to hear.

> A 55 year old woman was gardening with her husband one morning. She said to him 'What day is it — Monday?' He replied 'No, it is not Monday, it is Wednesday'. After a few moments she said 'Did I go to tennis?' 'No' he said, a little surprised, 'you did not go — it was raining'.
>
> Thirty seconds elapsed, then she said again 'What day is it — Monday?' This time he replied a little heatedly 'No, I told you, it is not Monday, it is Wednesday'. After a few moments she said 'Did I go to tennis?' 'No, you did not go — it was raining'.
>
> This sequence was repeated four or five times over the next few minutes. Having had a scientific training, the husband then began to systematically test his wife's memory. It emerged that she was unable to retain anything which he told her for more than a minute or so. She could not recall a visit by relatives the previous week or a holiday they had been on together some months before. Her memory of events 6 months or more prior to this episode was preserved. She spoke normally and behaved sensibly, apart from repeatedly asking the same question. He was becoming very alarmed when, after about 2 hours, she returned completely to normal.

The patient had the syndrome of transient global amnesia, a disorder which is thought to result from a temporary disturbance of function of both temporal lobes (and including the hippocampus). She asked the same question again and again because she forgot she had asked it and did not remember the reply.

Memory loss like this is not seen when function is impaired in only one temporal lobe; like hearing, memory is bilaterally represented. It is an important principle in clinical neurology that some functions can be carried out by either cerebral hemisphere and problems related to those functions may only be seen when both hemispheres are affected.

An experience common to us all is that distinctive smells can evoke a distant but poignant memory. It may come as no surprise that the temporal lobes are important for our sense of smell. The American neurologist Daly has provided a number of graphic descriptions of patients with epileptic fits arising in the temporal lobes. One of his patients had attacks that were 'preceded by the sensing of a vivid odor of peaches' followed by a feeling of 'I'm not really myself'. Frequently there was a sense of 'familiarity'.

Fits arising in the temporal lobe often involve smell and memory. When neurons in this lobe fire abnormally there may be an hallucination of smell (uncinate fit). There may also be a misperception of time whereby a new perception feels familiar, as though it has occurred before (deja vu).

## The Frontal Lobes

The frontal lobes (the cortical regions immediately anterior to the central sulcus) are important for motor control. Disorders of movement can be broadly categorized into problems which arise due to weakness, and those due to disorders of higher order motor control, by which we mean the planning, initiation, execution and termination of movement.

> A 50 year old woman suddenly became weak on the right side. She was unable to raise the right arm, the right side of the face drooped and she was unable to walk. A scan showed a large infarct in the left frontal lobe (F in figure 11.1).

The right sided weakness in this patient resulted from interruption of the motor pathway which runs from the left precentral gyrus (primary motor cortex) to the internal capsule. From there the tract crosses in the brainstem to the opposite side to innervate motoneurons in the spinal cord. These motor neurons than send signals via the nerves to the muscles, which ultimately give rise to movement (see sensory-motor chapter).

> A 70 year old woman complained of being unable to use her right hand. She was intelligent and sensible and there was no weakness of the hand. On being asked to make a fist and then to open it as quickly as possible, she was unable to do so immediately. After several seconds her hand opened very slowly. She also had great difficulty using the right hand to perform tasks such as undoing a button or using a type-writer. A PET scan showed impaired function in the medial part of the left frontal lobe.

This patient understood the command to open her right fist. She had normal strength in the hand. Her problem was an inability to initiate the movement. She had also lost the ability to execute skilled tasks.

Movement is a very complex process, many aspects of which remain poorly understood. Much of what is understood about motor control does not originate from lesion studies, but rather from observing the effects of stimulation of the motor areas. A somatotopically arranged motor area was first suggested by Hughlings Jackson in the 1860s, when he noticed that some patients with epileptic fits have twitching which begins in the fingers, marches up the arm, then to the face and finally to the body and legs. Such an arrangement was identified in monkeys by Ferrier in the 1870s, who used electrical stimulation of the open brain to produce movements. It was not until the 1950s that the existence of a 'homunculus' was finally confirmed in humans by the work of the neurosurgeon Penfield and his colleagues. During surgery performed on patients with epilepsy, these pioneers also used electrical stimulation of the exposed cortex to map the brain. In these experiments, the precentral gyrus was found to have the lowest motor threshold: stimulation produced simple, twitchlike movements. Most of the movements which resulted were contralateral to the side of stimulation, although bilateral movements of the oropharangeal, upper facial and truncal musculature were seen. Because of the nature of the movements and ease of stimulation, this region was termed primary motor cortex.

In these experiments, another somatotopic representation of the body was found on the medial surface of the brain anterior to the leg area of the primary motor cortex. Stimulation in this region produced movements of a different character: they were more complex, for example occurring across several joints, bilateral, or involving vocalization, and they often consisted of maintained contractions or postures. This region was named the supplementary motor area. As outlined in the sensory-motor chapter, the supplementary motor area and other non-primary motor areas are responsible for the many of the aspects of higher order motor control.

## Prefrontal Cortex

The frontal lobes account for one third the mass of the brain. Their relative prominence is one of the most distinct differences between the brains of humans and primates. We have considered the function of the part of the frontal lobes involved in the processing of movement. Much of the frontal lobe is comprised of cortex anterior to the motor areas, and is referred to as prefrontal cortex. Movement, perception, and intelligence can all be normal even with lesions of prefrontal cortex. Large lesions of prefrontal cortex can be clinically silent. Unlike other areas of cortex which we have discussed, circumscribed lesions of the prefrontal regions are not necessarily associated with specific cognitive deficits. What, then, is their function? Again, patients encountered in clinical practice offer some insights:

> A 50 year old man became withdrawn and unable to work. He only spoke when spoken to. Although he could still do things, the reasoning or purpose behind them seemed to be lost. On one occasion, his wife asked him to make her a cup of tea. Half an hour later, he was found in the kitchen surrounded by cups of tea, having used up the whole box of tea bags. In fact, once he began doing something, such as washing his hands, he often continued to do it repetitively. His behavior became unpredictable; frequently he left the table halfway through eating dinner to watch something which had happened to catch his eye on the television. He seemed no longer to care about how he appeared or behaved; he would say he needed to urinate, and proceed to do so in the living room. In the ward, he sat silently on his bed, endlessly fiddling with his pyjama cord. Then we realized he was compelled to use whatever object happened to be near his hands. When given a banana, he ate it, even though he wasn't hungry. Offered a pair of glasses, he put them on though they were not his own. On being shown a second pair, he put these on over the first pair. A third pair of glasses were also put on with some difficulty. His scan showed a large tumor involving both frontal lobes.

This patient demonstrates many of the features which are typical of patients with significant lesions of the prefrontal cortex. His personality changed: he repeated his actions inappropriately, a phenomenon called perseveration; his behavior became impulsive and disinhibited; his judgement became impaired. The behavior which we observed while he was in hospital showed that rather than being in control of himself, he was controlled by his environment. He would sit there doing nothing until a stimulus (or cue) came his way, whereupon he would engage in the behavior appropriate to that cue. This is called utilization behavior. In the process of becoming an adult, we acquire skills in using objects such as bicycles, pens, glasses etc. When these objects come before us, they cue the behavior normally associated with their use. In a normal brain, the response to these cues is sometimes suppressed. As a result of the frontal lobe tumor this patient was unable not to respond to every environmental cue which came his way. He could no longer plan or organize his own behavior.

As we have mentioned in a previous section, we are born with a repertoire of behavior patterns. These are seen in new born babies. An important function of the developing prefrontal cortex is to inhibit these behavior patterns.

This case illustrates some of the qualitative aspects of the function of prefrontal cortex. How can we summarize more precisely its role in our model of the brain? Luria regarded the frontal lobes as being responsible for the 'programming, regulation and verification of human activity'. Fuster wrote 'the prefrontal cortex supports several cognitive functions and … at least three of those functions can be identified as specific for that region of the neocortex: provisional (short-term) memory, preparatory set, and interference control. … all three functions cooperate under a supraordinate function that is distinctive of the prefrontal cortex as a whole: the temporal organization of behavior'. Damasio states that the frontal cortices 'select the responses that are most advantageous for an organism in a complex social environment'. Mesulam, in his review on neural networks, suggests that 'Through … widespread connections, the frontal lobes would be in a position to activate a given network, inhibit another, to influence network combinations, and perhaps even to allow internal readouts in a way that disengages the information processing from the response stage. The frontal lobes … would provide an arena for the various networks to play out different scenarios, the most successful of which may then dominate the landscape of neural activity'.

Perhaps the greatest insight into the role of prefrontal cortex comes from the knowledge that although patients with prefrontal lesions are sometimes able to score well in neuropsychological tests and a neurological examination, they fail miserably when dealing with the demands of life. Their knowledge and abilities are intact, but something is wrong with how they are applied. Exactly *what* signals or information are processed, and *how* they are processed, remain uncertain.

## The Basal ganglia

The precise functions of the basal ganglia, like prefrontal cortex, remain uncertain. However, it is clear from the effects of lesions in these structures that they are important in the modulation of movement.

> A 5 year old boy developed twisting movement of the left arm whenever he used it. Over the next five years his head began to turn and tilt to one side. Eventually the posture of all his limbs and trunk became distorted, even at rest. He was of normal intelligence. A scan showed no abnormality. A diagnosis of torsion dystonia was made.

Torsion dystonia is a disease which causes an abnormality of the balance of contraction between opposing muscles. As a result, the posture becomes distorted and there are abnormal movements. Many cases are hereditary. It is thought to be due to an abnormality of the basal ganglia, as patients with lesions in the basal ganglia can also develop these problems with movement. In torsion dystonia, however, it has never been possible to demonstrate any abnormality in the brain. With one or two exceptions, it is true to say that there is no exact relationship between structure and function in the basal ganglia.

> A 70 year old man presented with a 10 year history of Parkinson's disease. This had caused tremor, stiffness and slowness of movement. His major current problem was of difficulty in walking which he could only do very slowly, hesitantly and with frequent involuntary pauses. His wife happened to play a march on the record player one day and noticed that he walked much better when keeping time with the music. Amazingly, he was able to do intricate dance steps quite skillfully when the music was playing.

Parkinson's disease is associated with degeneration of the dopaminergic pathway which connects the substantia nigra to the basal ganglia (see the anatomy chapter). The range of symptoms varies markedly from patient to patient and it has never been possible to clearly relate a particular feature of the disease to a structural abnormality in the brain. In this patient a marked, if temporary, improvement occurred in his walking when he did this to music. This is an example of loss of higher control of movement which can be partially compensated for by using an external cue. It seems likely that spontaneous actions are processed differently from cued actions.

## A DISTRIBUTED MODEL OF NEURAL PROCESSING

The previous lesion-based journey through the brain considered neurological functions which appear to be processed in reasonably localized areas of the brain. The usefulness of lesion analysis in the practice of clinical neurology has led to an uncritical assumption — the notion that cognitive processes are processed independently in different parts of the brain. The analysis of disorders of language and memory has guided neurologists to the realization that processing of many neurological functions, rather than being strictly localized, is likely to be distributed.

### Language

Language is an incredibly flexible means of communication. That 'a picture paints a thousand words' is equally remarkable because of the thousand words, varying in form and meaning, which can be used to describe that picture. It can be argued that language is a localized function. In 95 per cent of right-handed and 70 per cent of left-handed people, language processing occurs in the left hemisphere. It is not understood why or how this relationship occurs. In most people, almost the whole of the right hemisphere can be destroyed with little effect on language capability, whereas a relatively small lesion in the left hemisphere can completely abolish the ability to communicate with language.

Two clinical syndromes suggest that language processing is further localized to specialized areas within the dominant (left) hemisphere. Bilateral lesions of Heschl's gyrus (primary auditory cortex), or unilateral left-sided lesions which also disconnect incoming pathways to the language area from the right primary auditory cortex, can cause a syndrome termed *word deafness*. Our patient in the previous section on the temporal lobes had one form of this condition. Such patients retain the ability to speak, name objects, read, and write. Often they can also identify non-verbal sounds, such as animal noises or music, and have normal pure tone audiometry results. However, the ability to comprehend spoken language and to repeat words is lost. Frequently, they no longer recognize speech as language, but describe it as an unpleasant buzzing or humming. Although such patients often have subtle abnormalities in language output, the disparity between comprehension and production of speech is dramatic. In contrast to word deafness, patients with *aphemia* (also known as pure word dumbness) lose their capacity to speak, with preservation of the ability to produce written language, comprehend spoken language, and read. Such patients usually have lesions of the left inferior frontal motor cortex and adjacent white matter. These two conditions demonstrate that well-localized lesions can lead to devastating but highly specialized loss of a component of language, whilst sparing other language functions.

What happens in language processing between the acts of decoding and formation of the sounds which make up speech?

The patient whom Broca described in 1861, although able to understand language without difficulty, could only say 'Ta'. Broca realized that this form of *aphasia* (acquired defect in language) was associated with lesions of the left inferior frontal lobe: this area of the brain has become known as Broca's area. Thirteen years later, the German neurologist Carl Wernicke described a different type of aphasia. Comprehension was severely impaired, and there was excessive speech production consisting of jumbled remnants of language. An example would be a patient who, when asked a simple question such as 'What did you have for breakfast?', replies 'The tooble win is Sunday, thank you'. He recognized this type of aphasia as being associated with lesions of the left temporoparietal region, now known as Wernicke's area.

For many years, it was thought that Wernicke's area contained all the components of language, somewhat like a dictionary, arranged as individual units, encoding all the information related to a particular word. For example, the unit representing 'dog' would encode concepts such as the four legs, wagging tail, fur and barking. As one popular neurology text put it, 'If the word is spoken, the pattern is transmitted from Wernicke's area to Broca's area and then on to the motor area that controls the speech muscles' (McLeod and Lance, 1983). Broca's area was thought to be responsible for arranging these words into sentences, and coordinating the oral movements required for articulation. This theory contained two assumptions about how language was processed. First, that it was strictly localized, with meaning coded in Wernicke's area, and expression coded in Broca's area. Second, it postulated serial processing: meaning was first elicited in Wernicke's area, and then transmitted to Broca's area if speech was required. Recently, this concept of language processing has been revised.

Damasio and Damasio (1989) conceive of three interacting networks which together allow language to operate. The first represents non-language interactions between the body and environment, or in other words concepts. The second processes purely linguistic functions such as recognition of phonemic combinations and syntax (phonemes are the most basic units of sound used in language, such as 'ba' or 'mu'; syntax refers to the grammatical structure of language). The third network mediates between the two. They cite as evidence the pathological conditions of achromatopsia, Wernicke's aphasia and color anomia, all of which can cause an inability to correctly name the color of an object.

Each of these deficits is associated with lesions in different sites in the brain. It emerges that a word and its associated concepts, rather than being encoded as one discrete anatomical unit, are coded in multiple regions of the brain. It can be concluded that language processing is therefore distributed rather than strictly localized. After hearing a word, what probably ensues is activation of a number of interconnected neurons, distributed widely in the brain, which together allow its correct interpretation and use.

In keeping with distributed function in the language areas, there are no structural, anatomical or physiological criteria which precisely define Broca's or Wernicke's areas. With the exception of word deafness and aphemia, correlation between language deficits and the anatomical location of lesions is imprecise.

Neither Broca's or Wernicke's aphasias are pure deficits of comprehension nor language output. This suggests that interpretation of meaning, word selection, and organization into correct syntax are not sequential processes taking place uniquely in either Broca's or Wernicke's area. It is not surprising, then, that there is neurophysiological evidence suggesting activation of these areas is simultaneous during language tasks. Furthermore, a number of structures outside these two areas, including the supplementary motor area and basal ganglia, are implicated in the processing of speech output, as evidenced from lesion studies. Therefore language comprehension and production is probably not a serial process, but instead involves parallel activation of multiple pathways.

Mesulam has used the term 'large scale neurocognitive networks' to describe neural systems involved in language and other cognitive processes. He postulates that they utilize parallel distributed processing with multiple iterative or reentrant circuits. The clinical, neuroanatomical and neurophysiological evidence is broadly in agreement with these theories.

## Memory

A full discussion of clinical disorders of memory is beyond the scope of this chapter. Kandel and Hawkins (1992) and Squire (1987) are more thorough reviews on this subject. In this section we will concentrate on evidence from lesion analysis to suggest that memory processing is both localized and distributed.

For neurological development and cognition to occur, neuronal interactions must be able to be sustained over time, be it milliseconds or decades. If this were not the case, processing would quickly disintegrate. Learning is the process by which new information is acquired, and memory refers to storage of that information so that it can be retrieved. Any model of the brain must account for learning and memory. Although precise definitions may differ, a distinction is recognized between short term memory and long term memory. Short term memory is what we use to retain and manipulate information required for a task immediately at hand. Information is stored only transiently and the facility has a limited capacity. Long term memory refers to information which is stored and then can be recalled after a longer delay.

In 1887, the Russian psychiatrist Korsakoff described a syndrome of severe amnesia which affects some alcoholic patients. It arises from deficiency of a basic vitamin, thiamine. Patients are left unable to lay down new memories. Stemming from his work, the critical importance of midline diencephalic structures such as the thalamus in the acquisition of new long term memories was discovered.

In the 1950s, the temporal lobes were recognized as the origin of some seizures. In the now infamous case of H.M., bilateral medial temporal lobectomies were performed in an attempt to alleviate intractable seizures. Following the operation, H.M. completely lost the ability to retain new information. He also lost some memory for previous events, with recently acquired memories being affected the most severely, and remotely acquired memories not affected at all. In other words, he was much like a patient with diencephalic amnesia. It has become apparent that the medial temporal lobes are vital for the acquisition and consolidation of certain types of memory. Further animal and human studies have identified the hippocampus and adjacent neocortex as being the important structures involved in these processes.

Patients with diencephalic and bilateral medial temporal lobe lesions lose the capacity to lay down a subtype of long term memory which is termed declarative memory. Such memories are recalled as a proposition or an image. Examples of this sort of memory include information from a passage in an encyclopaedia, the image of a painting, or events which occur during a person's life. The diencephalon and medial temporal lobes appear to be important not only in the acquisition of new knowledge, but also in recall of previously stored long term memories. Patients such as H.M. demonstrate a temporal gradient of retrograde memory loss. Recall of recently acquired long term memory is defective, whereas remote long term memory is recalled normally, suggesting retrieval of the latter does not rely on the integrity of these structures. There is no strict delineation between recent and remote long term memory: the degree of retrograde memory loss usually correlates with the degree of anterograde memory deficit, suggesting a relationship between learning and consolidation of memories.

In summary, focal lesions in these two areas can cause a severe and generalized deficit in the learning of new memories and retrieval of recent declarative memory. These processes appear to be localized functions reliant upon discrete structures in the brain.

We have mentioned that retrieval of remote long term memories is intact in patients with diencephalic and medial temporal lobe amnesias. Thus a patient might not be able to remember what happened two Christmases ago, and yet still be able to recall events from his childhood. A generalized deficit in recall of remote long term memories is not known to result from focal lesions. Why is this the case? One possibility is that recall of such memories is not a localized process, but one which is distributed. A corollary is that remote memories are also stored in a distributed network, rather than as a discrete trace in a localized area. A network model for short and long term memory has recently been postulated by Alvarez and Squire. This model proposes that the medial temporal lobe is vital for binding together connections between different areas of neocortex which are active in a newly acquired memory. It has a role in the storage and retrieval of recent long term memory. Over time, memories become consolidated as the links between different areas of neocortex active in a memory trace become independent of medial temporal lobe structures. The permanent repository of remote long term memories is therefore the neocortex.

Where are the individual components of the neural networks coding particular aspects of memory located? One possibility is the area of neocortex which also processes the cognitive modality to which that aspect relates (i.e. association cortex). For example, if asked to recall the sound of a train, we probably need to activate the visual association cortex to imagine what a train looks like, and the auditory association cortex to imagine what it sounds like. In support of this notion is the existence of modality-specific defects in cognition known as *agnosias*. For example, patients with prosopagnosia lose their ability to recognize familiar faces. In other words, such patients have lost their memory for faces. Their visual processing may otherwise be almost completely intact. They may have no difficulty in recognizing animals, objects or a painting. Furthermore, they will still be able to describe all the characteristics of a particular person when given a name. Prosopagnosia is most commonly associated with bilateral occipitotemporal lesions. Presumably neurons in these regions code the visual processing necessary to recognize people's faces, but not their other features. What remains uncertain is how the individual components of neural networks which support remote memories are coded, organized and activated.

Declarative memory is not the only neural system by which means information is stored. Procedural memory is measured as an improvement in performance or processing of a task (implying that neural interactions have been modified and retained) which occurs independent of awareness of recall. This capacity is preserved in patients with amnestic syndromes related to Korsakoff's syndrome or temporal lobectomies. For example, after multiple trials they can improve their performance in a maze task, yet are unable to learn the correct sequence from trial to trial, or even recall that they have performed the task on previous occasions. Although presumed to be related to pathways mediating sensorimotor integration, even less is known about the networks mediating procedural memory than those which subserve declarative memory.

The existence of procedural memory raises an interesting point: that memory might be an intrinsic quality of neural interactions. The study of patients with memory disorders can provide insight into the nature of memory. While these observations have suggested plausible pathways and broad mechanisms which function to support memory, the neuronal or synaptic mechanisms that operate remain unknown (although neuroscientists have discovered mechanisms such as long term potentiation [LTP] which show potential as a mechanism underlying aspects of memory).

## MIND VERSUS BRAIN — INSIGHTS FROM NEUROLOGY

We have learnt about how neurologists have come to the realization that some brain functions are processed in distinct areas of the brain, whereas others appear to be diffusely represented. A distinctive characteristic of the brain is that neurological functions are integrated. We are only aware of the anatomical and functional separations which we have made if something goes wrong. This has led to a philosophical, if not scientific, separation between the brain, and its emergent functional properties, or in other words the mind. Many investigators regard the concept of the mind as being synonymous with consciousness (see chapter 4, The

Mind–Brain problem).

Firstly, activity and arousal do not equate to normal consciousness. Spinal mechanisms are enough to mediate some movements. For example, in paraplegics in whom the spinal cord has been completely transected, stimulation of the skin can give rise to withdrawal of the lower limbs, even though no input from the brain can reach the segment of the spinal cord mediating those movements. In the persistent vegetative state, which results from severe, diffuse injury to the cortex with or without subcortical damage, patients may have sleep-wake cycles, the eyes may be open and gaze about the room, yet there is no evidence of awareness of either the environment or self. The neural substrate for arousal, which is the reticular activating system, appears not to be sufficient to mediate awareness.

Even with normal arousal, awareness, and attention, consciousness is not necessarily a unified entity. Geschwind (1965) made this observation when discussing the case of callosal apraxia, described in collaboration with Kaplan. Their patient suffered disconnection of the language area in the left hemisphere from the sensorimotor cortex in the right hemisphere. His ability to communicate with language was completely intact. He could therefore describe how to use a hammer. Asked to show how he would use a hammer with his left hand, he was unable to do so. Yet when a hammer was placed in his left hand, he could demonstrate its use correctly, without being able to name it. Geschwind summarized his interpretation in the following manner: 'that part of the patient that could speak normally was not the same part of the patient which 'knew' (non-verbally) what was in the left hand'. Non-verbal awareness in the right hemisphere and verbal awareness in the left hemisphere existed independently, each unable to access the other in order to form a normal integrated response.

Awareness, defined as responsiveness to the environment, can also exist independently of arousal. During stage 2 sleep, it has been shown on an electro-encephalograph (EEG) that brain electrical activity can change in response to a stimulus, without any evidence of clinical arousal, and without further change in the level of sleep as determined by the EEG. If the stimulus continues, such as a series of clicks, the response quickly extinguishes. However, if a person's name is called instead of continuing the clicks, the response can reappear.

Therefore consciousness is not an all or nothing phenomenon. Consciousness, on the basis of observations made in brain-injured patients, appears to be comprised of overlapping functions which can, on rare occasions, be damaged selectively.

Perhaps the most significant assumption in mind–body relations is that the mind, manifested as consciousness, is hierarchically superior to the body. In other words, that we possess free will, at least over conscious motor acts. We return here to the frontal lobe syndrome of utilization behavior. An observation in one of our patients provides some insight into this dilemma. This patient was shown not to be aphasic by testing with yes/no questions and simple commands. Utilization behavior was elicited by placing a jug of water and empty glass near his hands: he poured the water into the glass. We then said to him 'Don't pour the water into the glass', and he confirmed that he understood us. Utilization behavior was again successfully elicited in the same manner (he poured the water into the glass). We then asked 'What did we tell you to do?' and he replied 'Don't pour the water into the glass'. When asked 'Why did you do that then, pour the water into the glass?' he was silent and looked perplexed and uncomfortable, as if not comprehending his own behavior.

Finally, a study has been performed to examine the time at which subjects are first aware of their own decision to perform a spontaneous motor act. Electrical activity in the brain can be shown to change 350 milliseconds *prior* to the moment that subjects identify as the time when they made that decision. Such a delay does not occur between the time that sensory signals reach the cortex and awareness of those signals. These results highlight the subtleties of consciousness, but also indicate that consciousness of the decision to perform a motor act follows rather than precedes the events in the brain leading to that movement.

## TOWARDS INTEGRATED MODELS OF BRAIN FUNCTION

Progress in understanding the brain as an integrated system has advanced slowly in neurological circles. Over two millenia passed before it was firmly recognized that cognitive functions were processed by the substance of the brain. In the last two hundred years, neurologists have focused on the attempt to localize different motor, sensory and cognitive functions to specific regions in the brain. This pursuit has been fuelled by the success of the localizationist model in specific clinical settings. However, the limitations of this model in other instances has always been evident.

Insights provided by neurologists and lesion studies still seem likely to contribute significantly to multidisciplinary neuroscience, albeit to elucidate the co-existence of localizationist and distributed models, and highlight the need for models that integrate them.

## REFERENCES AND SUGGESTED FURTHER READING

Chugani H. T, Phelps M. E, Mazziotta J. C. (1987) Positron emission tomography study of human brain functional development. *Ann Neurol*, **22**, 487–97.

Damasio H., and Damasio A.R. (1989) *Lesion Analysis in Neuropsychology*. New York, Oxford University Press.

Fuster J.M. (1989) The prefrontal cortex: anatomy, physiology and neuropsychology of the frontal lobe. New York: Raven Press.

Geschwind N. (1965) Disconnexion syndromes in animals and man, parts I and II. *Brain*, **88**, 237–94, 585–644.

Jackson J. H. (1880) On affectations of speech from diseases of the brain. *Brain*, **2**, 203–22, 323–56.

Kandel E. R, Hawkins R. D. (1992) The biological basis of learning and individuality. *Scientific American*, September: 53–60.

Lashsley K. S. (1963) *Brain Mechanisms and Intelligence*. A Quantitative Study of Injuries to the Brain. 1st ed. New York, Hafner.

Luria AR. (1973) *The Working Brain: An Introduction to Neuropsychology*. 1st ed. Middlesex, Penguin.

Mesulam M. M. (1998) From sensation to cognition. *Brain*, **121**,1013–52.

Plum F, Posner J. B. (1982) *The Diagnosis of Stupor and Coma*. 3rd ed. Philadelphia, FA Davis.

Rumelhart D. E, McClelland J. L. D. (1986) Research Group, eds. *Parallel Distributed Processing: Explorations in the Microstructure of Cognition*. Volume 1: Foundations. 1st ed. Cambridge, MIT Press.

Spillane J. D. (1981) *The Doctrine of the Nerves*. 1st ed. Oxford: Oxford University Press.

Squire L. R. (1987) *Memory and brain*. 1st ed. New York: Oxford University Press.

Walsh E. R. (1993) *Clinical Neuropsychology*. 3rd ed. New York: Oxford University Press.

# Chapter 12

# MODELS OF THE BRAIN IN PSYCHOLOGY

Iain S McGregor[1] and William C Schmidt[2]

[1]Department of Psychology, University of Sydney
[2]Department of Psychology, Dalhousie University

An old neuroscientific saying states that 'The human brain is the most complex structure in the known universe and is also who you are'. A moment's reflection on this adage brings home the immense complexity of the task facing the would-be brain modeler. Despite the profound difficulties encountered in studying such an elusive topic, some understanding of the specific mechanisms and principles behind the human brain and mind is starting to emerge. Since the time of Descartes, thinking has been dominated by the role of the brain in giving rise to thought, remembering, feeling and other aspects of human behavior.

In psychology, a systematic experimental approach is adopted in the hope of extending the philosophy of mind into an empirical science. Because the questions that can be addressed by any one experiment are limited, linking experimental outcomes back to the bigger questions posed within philosophy becomes a taxing endeavor. Regardless of the difficulty of empirically testing theories of the mind, the interpretation of experimental results is often founded on clearly identifiable philosophical positions and approaches. Reflecting these varied approaches, experimental psychology can be divided into a collection of over thirty subdisciplines, each devoted to understanding specific aspects of the mind and behavior.

Given the degree of specialization that has taken place in psychology it is not surprising to find that there is great variety in the way that psychologists view the human brain, and in the degree to which psychologists expect that directly studying the organ will lead to an understanding of the human mind. Some observers fear that an overly reductionistic psychology which only views mind in terms of brain is in danger of collapsing the discipline of psychology as a distinct entity that will render psychology little more than a sub-discipline of biology. Such critics might claim that we may be able to understand how the brain works and is structured, but this is no guarantee that we will understand how it gives rise to the human mind. Strong advocates of this view suggest that study of the mind can proceed independently of knowledge of the brain. Opposing views see a solid foundation in the neurosciences as being the only way ahead for contemporary psychology, in that explanations of mental life that pay no attention to brain processes are in some ways always bound to be little more than inspired guesses at what the biological reality may be (Edelman, 1992). As Watts (1992) somewhat pessimistically remarks on this tension within psychology: 'the sad fact is that the various sub-disciplines of psychology have never really had much more than a marriage of convenience. Now, alternative partners are increasingly available, and the weakness of the marriage becomes transparent' (pp. 490).

## SOME OF THE AREAS OF PSYCHOLOGICAL INQUIRY

There are several subdisciplines of psychology that concern themselves directly with how the brain functions. Behavioral Neuroscience is a subdiscipline whose scientists probe the brain to assess its role in the control of behavior. These scientists typically work with animals, often rodents, with the assumption that understanding brain function in a sub-primate species will yield fundamental information concerning the structure and function of the human brain. In the course of their work, these researchers use methodologies such as the electrical and chemical stimulation of the brain, assessment of the behavioral effects of brain damage and observations of changes in brain function resulting from various stimuli. In the past three decades a huge database of information has been amassed concerning the rodent brain and its control of behavior, from which inferences have been drawn about human brain function.

Psychophysiology is a subdiscipline involved in probing brain function through examination of brain and body function, with measures including heart, respiratory and sweat rate, and electrical brain function. Cognitive Neuroscience is a related discipline which studies higher intellectual function via the use of an array of sophisticated computerized imaging techniques (see chapter 15).

Many psychology departments will employ a Psychopharmacologist, to examine the effects of drugs on the brain and behavior (using either animal models or human subjects). Although not always directly studying the brain, the psychopharmacologist will usually offer receptor level explanations for observed behavioral effects.

Most psychology departments will also house a Neuropsychologist, to assess the behavioral and cognitive changes occurring as a result of brain damage (lesion–behavior relationships) in humans.

Many other areas of psychology are not as concerned with directly studying the brain, but wish to understand the mind and mental processes nonetheless. Workers in the area of Cognitive Psychology endeavor to understand how people think — both in terms of how human subjects consciously think and reason about specific problems, and how they unintentionally operate in carrying out everyday tasks. However, as we will see, these scientists do not always believe that an understanding of how the brain functions is necessary for us to understand how higher order cognitive functions operate (also see chapter 4, The Mind–Brain problem and chapter 9, Computer Models of the Brain).

There is a fine line between perception and cognition. Perceptual psychologists and psychophysicists endeavor to understand how the human perceptual apparatus processes information from the world, before it is used in the execution of a specific task. Psychophysics results can also shed light on the mechanisms of the brain that are used to perceive the world.

We outline three approaches to model the brain, and then present models of five specific psychological functions.

## PSYCHOLOGY-BASED APPROACHES TO MODEL THE BRAIN
### Localization of function

Playing a large role in approaches to studying the brain has been the complex questions of whether psychological functions map directly onto brain structures in a one-to-one fashion (i.e. are spatially localized within the brain), or whether numerous brain areas co-operate in order to implement a given psychological function (i.e. processing is parallel and spatially distributed).

The psychological debates surrounding localization of function in the brain continue across a range of disciplines to this time (for example, see the previous chapter), and much of the history of ideas in these disciplines, overlap.

The methodological objections leveled at the proponents of localization did little to hinder their progress. The work of Sir Charles Sherrington in the early 20th century focused on the motor cortex, and half a century later, the work of Dr Wilder Penfield and colleagues made the localizationist approach even more respectable. Penfield used systematic electrical stimulation to probe the brains of conscious epileptic patients who were being prepared for surgery. Penfield's explorations led to the discovery that the entire surface of the body is represented spatially on the surface of the somatosensory cortex. Penfield's data led to the construction of the famed 'homunculus', a cartoon representation of a human with all of its body parts drawn to scale according to the relative proportion of the sensory-motor cortex that each occupied (see figure 1.2 in The Big Picture).

The methods of localization were criticized again early in the 20th century when a skeptical holist named Karl Lashley empirically studied the technique of ablation. Lashley demonstrated that the degree of damage to maze learning behavior in rats depended on the amount of tissue removed rather than on its exact location. This finding led him to propose the concepts of 'mass-action and equipotentiality'. Lashley's idea of equipotentiality suggested that any functional area of the brain had an equivalent potential or capacity to carry out a particular behavior. This implied that neural processes are plastic, and Lashley introduced the idea that nearby areas of the brain might adapt to take over the functioning that occurs in a damaged region.

Lashley's law of mass action suggested that the efficiency of performance of a given function may be reduced according to the extent of brain injury within any one area. According to Lashley, the destruction of larger pockets of tissue results in the greater likelihood of affecting function. Lashley conceptualized behavior as arising from the spatially distributed parallel processing of a group of neurons in the brain. Because each of these neurons contributes to causing behavior, the destruction of any single area of the brain results in the hampering of a wide variety of functions. However, damage limited to a single location does not necessarily result in the destruction of any one function because the processing mechanism is distributed throughout many locations. A single punctuate lesion is unlikely to completely wipe out any one process, yet will affect many others.

## Early Connectionism and Hebbian learning rules

By the middle of the 20th century, both the localizationist and holist proponents had presented convincing data and arguments with little synthesis of these positions. At about this time, Donald O. Hebb, a student of Lashley's, published *The Organization of Behavior* (1949). In this volume, Hebb argued that the behavioral repertoire is built up gradually from a large number of discrete, yet spatially distinct neuronal processing units that he called 'cell assemblies'. Because of the limited processing capacity of cell assemblies, simple behavioral functions or sensations could be localized within specific regions of the brain. More complex behaviors were hypothesized to require more processing power, and to be generated from the connection of a number of such assemblies. This organization made the origins of complex behaviors less localizable and less amenable to breakdown from punctuate lesions.

Hebb provided a learning rule, through which groups of neurons could become associated over time. Hebb's rule postulated that if two spatially adjacent cell assemblies were simultaneously active, then the strength of connectivity between them increased. The more strongly connected two cell assemblies, the more likely that one would become activated if the other was active. This simple learning rule would account for the firing of cell assemblies that were related, resulting in what Hebb called a phase sequence: a sequential pattern of firing cell assemblies. Hebb postulated that behavioral patterns are built up gradually as cell assemblies become interconnected through experience with the world. Notice that the mechanisms of phase sequences need not be localized in a particular spatial region or location since the pathway of firing could just as easily lead away from as towards a given neural center. At the same time however, the theory suggests that cell assemblies required for similar functions and content (e.g. different aspects of vision) will form close to their similar input structures, resulting in local regions of cell assemblies concerned with processing similar types of stimuli, while dissimilar material will be stored at more diverse locations. Hebb's framework holds the potential to accommodate the holist and the localizationist. In addition, it was suitable to explain observations of neural plasticity and the recovery of function after brain damage.

Hebb's work at synthesizing the holist and localizationist approaches contributed greatly to the foundation of neuropsychology as a separate discipline. Since Hebb's time, research into Hebbian learning has continued, particularly with the recent computational resurgence of Connectionism (see chapter on computer models). Diverse topic areas from neuropsychological processes to language are amenable to modeling using Connectionism methods.

## Modularity: Horizontal versus Vertical Processes

The notion of compartmentalization of mental function, has matured through several distinct stages. Crucial to each of these stages is the notion that definable component processes cooperate to produce mental phenomena.

The philosopher Jerry Fodor (Fodor, 1983) has been responsible for a resurgence of this approach. He relies heavily on the distinction between 'horizontal' and 'vertical' processes. Vertical processes execute domain specific tasks: tasks that are well defined and do not require knowledge outside of the narrow informational matter that they were designed to deal with. In contrast, horizontal processes deal with information from a diverse set of topic areas and operate across a large number of domains. This distinction is quite useful because it can help us to think about what sorts of processes we can expect to be highly localized in the brain, and what sorts of processes probably are not. Since vertical processes are quite limited in their scope of function, and since they deal with a restricted amount of information, it is reasonable to hypothesize that they might be self-contained processing units and therefore highly localizable. Similarly, because the scope of horizontal processes is far reaching and requires informational input from, and access to, a large number of different domains, it is reasonable to suspect that such processes are spatially diverse and less easily localized.

Fodor's taxonomy distinguishes sensory transducers, input analyzers, and a central system as the major components involved in the mind and brain. Sensory transducers simply change energy in the world into neural firing. The input analyzers are vertical processes that transform and recode the transducer's neural firing into a format that can later be used by the horizontal central processes. Fodor suggests that the input analyzers are 'modular', signifying their operational independence and are self-contained units, both physically and computationally, and associated with a fixed neural architecture making them highly localizable in the brain. These modular systems are informationally encapsulated. This means that the processing structure contains all of the information that it needs to perform its designated computation, and it requires access to no other information (particularly information from higher level, cognitive functions) in order to perform its task. Informational encapsulation guarantees that the function is a vertical process, and it is a necessary feature of modular processes.

In contrast to modular processes, central systems are not domain specific, do not operate in a mandatory fashion, are slower, and are distributed throughout the brain. It should be noted that very little is known about the mechanisms of how these central systems operate, and they are labeled as such largely as a convenient way of referring to the collective of horizontal processes.

If Fodor's modularity hypothesis is correct, then the best scientific approach to analyzing brain function is first to investigate modular processes in order to provide the functional architecture upon which cognition is built. After enumerating a large number of such processes, the task of determining how they provide a basis for the more elusive and mysterious central systems can be more easily approached. It is no coincidence that an elucidation of Fodor's modular framework happens to be the current direction that the cognitive sciences have been taking. By restricting the scope of the problem, the solution becomes more manageable.

Modelers of brain function have also taken a modular approach to understanding the interaction between form and function. Just as artificial intelligence researchers have discovered that they have to restrict their problem solving to microworlds, so too have psychological modelers.

## SPECIFIC BRAIN MODELS OF PSYCHOLOGICAL FUNCTION

We will now turn to some implemented psychological models that present contemporary approaches to understanding systems involved in the mind's overall cognitive functioning. For the most part, the processes discussed can be thought of as vertical modular processes, but they are often instances of a class of horizontal processes. For instance, while a model of memory is horizontal, models of iconic memory, auditory memory, or face memory are more likely to be vertical. Similarly, while visual processing is horizontal, components of it seem to be vertical (i.e. face, orientation or color processing).

We present five examples of models of brain function from a psychological perspective.

## 1. VISUAL PERCEPTION

A wealth of detail has emerged about the neural mechanisms underlying visual perception. These processes illustrate general principles that have found favor in modeling efforts within psychology. The visual system is perhaps the best understood (although one of the most complex) perceptual systems in the brain. Estimates suggest that as much as 70% of the normal brain carries out vision-related processing. At the periphery of this system are sensory receptors (in the retina). These peripheral processing units feed into input analyzers which in turn transmit data to, and transform data in, a location at the back of the brain where the primary visual cortex is housed in the occipital lobe (see figure 1.2 in The Big Picture). This brain center has been decomposed into a large number of modules (also input analyzers) which carry out specific functions.

In general, there is a highly structured division of labor between different cortical regions involved in processing different types of visual information. The primary visual cortex performs elementary feature detection on its inputs and acts as a distribution center, where incoming information is sent to further modules located in the extrastriate cortex for processing of color, motion, and depth, and the temporal cortex for form-based object recognition and the posterior parietal cortex for visual attention and spatial orientation.

Such models in psychology, in some ways overlap with the sensory model in the sensory-motor chapter (for example, both provide details of processing within the visual system as modular in structure), but the psychological models emphasize the parallel and distributed processes. There are multiple functional modules within the visual system that do not necessarily act in a sequential or hierarchical manner, but rather act in a simultaneous and integrated fashion based on dense interconnectivity. The notion of reentrant connections is of primary importance, whereby different functional modules display massive parallel and reciprocal anatomical connections, presumably allowing integration of distributed information. This parallel distributed processing probably explains how the brain can derive a response based on multiple aspects of sensory input in such a short period of time — the processing task is farmed out to a large number of local decision makers whose outputs are then integrated as a whole.

### The Binding Problem

The principles of modularity and parallel distributed processing lead to a question of particular interest to psychologists. How do the activities of many separate processing modules combine to give us a coherent, unified and familiar picture of the world? How do we put it all together? This issue is known as the binding problem. Basically we don't know the answer to this question yet. In fact, some researchers dispute that it is even a problem (experience may just appear introspectively to be a single, fused event). One exemplar model for solving the binding problem is the synchronous processing model.

The synchronous processing model of Singer and colleagues (Singer and Gray, 1995), suggests that during the processing of a visual stimulus, interconnected neurons in different visual cortical regions have a strong tendency to fire in synchrony with oscillating burst patterns at Gamma frequencies between 30–70 Hz (see further Gamma information in The Big Picture, chapter 10b and figure 15.2) . It is proposed that the population of responding neurons (representing their associated referents in the environment) being bound together are distinguished through a temporal (timing) code. Through the use of single unit neuron recording techniques, evidence for synchronous firing to related environmental objects (time locked within temporal periods as short as a few milliseconds) has been found both within and between hemispheres, and across different modules within the visual system. It has been proposed therefore, that the information about the precise location of a single stimulus, and about that stimulus' attributes, is not represented solely in the response of a few activated neurons, but is encoded in the graded responses of the aggregate of neurons that fire to a particular stimulus. One effect that the distributed encoding and processing proposed by the synchronous processing model has, is that it drastically reduces the number of neurons required to represent different patterns of environmental stimulation, and it fits naturally with the types of graceful degradation that are often observed with damaged neural systems.

True to Hebb's cell assembly theory which acted as a foundation for these ideas, a single neuron is likely to participate at different times in the representations of different sensory and motor patterns. This makes destroying the entire network by selectively damaging localized areas of it difficult. Furthermore, the Hebbian learning mechanism (the strengthening of connections that are co-activated) supplies insight into how such synchronous processing structures could be constructed through experience. When an environmental source of input appears, it mechanistically elicits responses in the most peripheral neural systems which propagate to more richly connected cell assemblies in a fashion time-locked with the input. Because the cell assembly apparatus is mechanical, a similar stimulus-controlled pattern of firing would be expected to unfold in response to similar inputs.

The idea that correlated bursts of firing, performing a variety of computations in different modules, are conjoined through temporal coincidence to yield a single percept of the world, is intriguing. However, despite mustering a fair bit of support at the neuronal level, much more work is necessary to investigate this model of perception.

The binding problem demonstrates how in considering visual perception, psychologists and neuroscientists have had to use the principles of localization of function, selectivism, modularity of processing, parallel distributed processing and synchronous processing to explain seemingly simple tasks that we effortlessly carry out everyday. As this example illustrates, science has stumbled onto the requirement of these concepts through the gradual process of refining theories of the brain's operation to fit more and more of the data. Back to the overall model of visual perception.

## Separate Visual Pathways for Perception and Action

Another example of the scientific approach to understanding the visual brain demonstrates how many of the previously discussed brain concepts are put into action. A specific example of some rudimentary input analyzers are X, Y and W ganglion cells of the retina. These cells receive information directly form retinotopic photoreceptors, and they transform this information and relay it to other visual centers for elaborative processing. X cells sustain their response for as long as visual stimulation are present, have relatively small receptive fields, and their axons show medium conduction velocities. Y cells, on the other hand, respond transiently to visual stimuli (primarily fire upon a stimulus' appearance or disappearance), have large receptive fields and fast conduction velocities. W cells have the slowest conducting axons of the cell types discussed here, and show a broad range of different response characteristics. The processing distinction between these three types of cell appears to be at least partially preserved in terms of the structure of the lateral geniculate nucleus of the thalamus (L in figure 1.2), that is responsible for transmitting information from the eye to the occipital lobes (and elsewhere). Further evidence shows that even at the level of the striate visual cortex, a distinction can be made between visual modules that receive primarily X cell input (the parvocellular stream) or Y cell input (the magnocellular stream).

Magnocellular systems process transients, motion and are relatively impaired at color, while the opposite primarily holds (with exceptions) for parvocellular systems. Some have held that the parvocellular stream is primarily responsible for computing object properties (perceptual properties such as color, identity, object relations, etc.) while the magnocellular stream is responsible for computing location information about objects (i.e. movement, distance away from the viewer, motion trajectories, etc.; Mishkin, Ungerleider and Macko, 1983). This hypothesis, which would see the visual processing stream segregated into primarily a 'what' processing stream (parvocellular) versus a 'where' (magnocellular) stream, has been expanded and applied to human performance data by numerous researchers, and a thorough coverage of this topic can be found in van der Heijden (van der Heijden, 1992).

Livingstone and Hubel (Livingstone and Hubel, 1988) have anatomically extended the distinction between what and where processing streams. They proposed that this rough taxonomy of processing (what versus where, parvocellular versus magnocellular) extended well beyond the primary visual cortex, with the parvocellular system mapping onto processes located ventrally in the brain (performing functions such as object recognition), and the magnocellular stream feeding primarily into dorsally located brain centers. Unfortunately, nature has not been shown to be this neat, with much evidence of cross-talk even between the magnocellular and parvocellular systems having been recently discovered (see Milner and Goodale, 1995).

A recently emerging model of brain organization that has been influencing the direction of research in both neuropsychology and cognitive psychology, is informally known as the two visual systems hypothesis. Like the original Ungerleider and Mishkin (1982) proposal, this model capitalizes on the early magnocellular versus parvocellular processing distinction. However, it encourages us to consider the tasks for which these brain structures have evolved as evidence towards the types of functions that these structures are likely to implement. One pathway, the ventral-parietal pathway, has been identified with processing location information of important events. The other pathway, the dorsal-temporal pathway, has been associated with the processing of qualitative, perceptual information about the same events. More recently, Goodale and Milner (Milner and Goodale, 1995) have proposed that these pathways might be conceptualized as processing 'action towards objects' versus 'object recognition' as opposed simply to 'location' versus 'feature'. Casting these processing structures in this light suggests why cross-talk would be expected to exist between magnocellular and parvocellular systems.

The importance of this research line (as with its like-minded predecessors) is that the two visual systems hypothesis makes direct, testable predictions about behavior (function) based on the structure of the brain. Furthermore, it presents an opportunity to examine often illusive structure to function correspondences.

As an interesting illustration of how these brain structures can affect behavioral performance, the pathways can be separated during a saccadic suppression task (Bridgeman et al., 1981). When people make quick eye movements (saccades), there is a period of time during which the input to the visual system is suppressed. Various psychologists have focused on this period and displaced the visual target, in order to see how the observer will respond to the altered visual world. The observers saccade to the location that the target appeared during saccade initiation, and then correct their gaze after the saccade has been executed. Cognitively, observers are unaware of the change, yet if asked to simultaneously point to the target location, they point to the location of the displaced target. This means that the processes coordinating pointing action indexed the location of the displaced target, while the saccadic programming and perceptual cognitive processes failed to have knowledge of the changed information. This suggests separate processing pathways involved for action and perception.

The idea that multiple parallel processing pathways exist is not new. The notion that the action that the individual plans to perform acts to select the processing pathway that has control over action is rather novel. This possibility has strong consequence for psychology's approach to studying behavior. It suggests that examining the task demands will be more important than we have considered it to be in the past. Not only must the environmental input and the mental set of the organism be considered, but the task demands and the set of possible processes that could respond in parallel to the input must be considered when attempting to explain behavior. Even in simple tasks, it may not be safe to take for granted that given a certain input condition, the same response will systematically occur.

The program of research evaluating the model of separate brain processing for perception and action is important on a number of fronts. Firstly, it assists in delineating the type of modular processes that exist. Secondly, it demonstrates specific structure to function correspondences between brain and behavior. Finally, it is beginning to link a number of distinct approaches to understanding brain function in psychology. Research from anatomy, neuropsychology, perception, and cognitive psychology is beginning to be integrated.

## 2. ATTENTION

'Attention', William James said, 'is the taking possession by the mind, in a clear and vivid form, of one out of what seem several simultaneously possible objects or trains of thought'. What kind of processes contribute to this ability to select information from the environment for focal (conscious) processing? Whereas visual processing seems to be primarily localized in the occipital lobe, centers involved in attentional processing are spread throughout the brain. In other words, vision is primarily a vertical faculty with processing at one stage feeding into processing at others, while attention appears to be a horizontal faculty with its effects either occurring within other modules, or within a global system that modulates other modules.

As with visual perception, attention appears to involve a large network of modular processes involving many localizable brain regions, often acting in unison. Unlike the visual system, attentional processing systems are widespread and distributed throughout the brain. This is illustrated by the fact that damage to a variety of different cortical and subcortical regions can lead to quite similar deficits in attention.

A network for visual attention has been modeled by the American neurologist Marcel Mesulam (Mesulam, 1990). Mesulam's network involves at least three 'neurocognitive maps' corresponding to cortical regions where damage produces attentional deficits. Firstly, there is a sensory map based in the posterior parietal cortex — working in conjunction with the midbrain and thalamic structures to shift our visual attention from one object to another. Secondly, a motor map based in the frontal eye fields is involved in eye movements, scanning patterns and reaching and grasping for attended objects. Thirdly, a motivational map mediated by the anterior cingulate gyrus ensures that attention is switched toward objects with the greatest motivational significance. Many neuroscientists have noted that the anterior cingulate gyrus seems to play an executive role in cognition, guiding attention during complex cognitive and verbal tasks or in situations where attention to one of many complex stimulus attributes must be made. According to Mesulam, directed attention involves the simultaneous functioning of all three cortical regions in a co-operative manner. Activation of this network is proposed to be regulated by the locus ceruleus in the brainstem which modulates the ongoing level of arousal.

The Mesulam model bears certain anatomical, if not conceptual, similarity to a competing model of attention proposed by Posner and Peterson (1990) outlined in The Big Picture (figure 1.6). They have proposed that there are at least three independent anatomical networks involved in attention, and they have identified an orienting network, an executive network, and an alerting network.

The orienting network is tightly connected to the brain centers responsible for eye movements, and it is proposed that attentional facilitation of a visual spatial region precedes the movement of the eyes to that region. Facilitated processing of information at this destination occurs for 50–150 milliseconds after attention has been shifted there. Three brain areas are postulated to be involved in this system: the posterior parietal lobe, the superior colliculus, and the pulvinar. Damage to any one of these areas results in the reduced ability to shift attention without eye movements, yet damage to each area results in characteristic deficits, thereby emphasizing that each area is performing a specific role in attentional orienting. Damage to the posterior parietal lobe affects the subject's ability to disengage attention from its current location. Damage to the superior colliculus affects the way that movement of attention is actually undertaken, and damage to the midbrain (including the pulvinar) has shown characteristic ignoring of information contralateral to the locus of the damage.

One of the hallmarks of attention is that there are limits in the number of locations that can be attended at any one time. Even when two targets to be detected in different modalities (i.e. auditory and visual), there is interference from one modality with performance in the other. Posner takes this as evidence of the operation of a unified component to the attentional system. There is evidence that areas of the midfrontal lobe, especially the anterior cingulate gyrus, are involved in an executive attentional network. This network has been found to be active during selective visual attentional tasks as well as language tasks, and has inputs from areas that when damaged, result in a general degrading of general attentional performance.

The final attentional network identified by Posner is the alerting network, whose function is proposed as maintaining an alert, involved state. This network contributes to the formation of two hallmark qualities associated with attention: increasing accuracy and reducing response time. Patients with lesions in the right frontal area have difficulty maintaining an alert state, and blood flow studies during the execution of tasks requiring sustained vigilance have revealed the activation of right frontal lobe and parietal areas. The involvement of the neurotransmitter norepinephrine appears to be of importance here. Pathways arising in the midbrain are cortically distributed via the right frontal lobe. Studies using selective blockers of norepinephrine eliminate the normal effect that warning signals have on performance (the reduction of reaction time).

The summaries of the models introduced in the current section illustrate that models of attention must be very complex, in that they have to account for local changes in attention as well as the distributed networks as described in the Mesulam and Posner models. Posner, who examines covert orienting in visual attention, goes beyond Mesulam by proposing that the areas highlighted by Mesulam, the anterior cingulate gyrus and the posterior parietal cortex are able to implement the amplification effects seen in specific brain regions during attention to specific stimulus attributes. Precisely how this is achieved is one of the greater puzzles to be faced by future models of attentional processes.

## 3. LANGUAGE

The study of brain functions involved in language provide an interesting focus for the localizationist model of brain function (recall that the language platform was largely the reason that holist ideas lost favor in the late 19th century). As mentioned in the chapter on neurology, Paul Broca demonstrated through autopsy that an impairment in speech production (known as aphasia — a general term used to denote the disruption of speech) followed damage to a restricted part of the left frontal lobe, now known as Broca's area. Carl Wernicke added support to this localizationist model when he distinguished a second sort of aphasia, which had to do only with speech comprehension, and which resulted from damage to a specific region of the superior temporal lobe, now known as Wernicke's area. The notion of lateralization of function has also been prominent in localizationist models of language with the overwhelming majority of people having speech production and comprehension mechanisms primarily localized in the left cerebral hemisphere.

Today, the importance of Broca and Wernicke's areas in language is still apparent, although recent theory has moved beyond this simple dual center model, and as with the perception and attentional models, has seen language as emerging from the parallel operation of many different brain regions, each having different functional specialization. This new distributed model is associated with investigators such as Marcus Raichle (1994) and has been largely driven by brain imaging measures of humans engaged in various aspects of language.

For example, listening to a word being spoken activates the primary auditory cortex, Wernicke's area and the left posterior temporal-parietal cortex. Reading nouns without speaking them activates the primary visual cortex and the medial extrastriate cortex. Interestingly, the left medial extrastriate cortex seems to be only activated by actual word-like forms rather than by random strings of letters, suggesting it to be a word form recognition area. Repeating a word aloud activates motor as well as sensory areas, including the insula, paramedian cerebellum, primary motor cortex and supplementary motor area.

Creative processes in language production seem to activate still other centers. When subjects are asked to generate a verb to a visually presented noun (e.g. the word 'cake' is presented and you say 'eat'), the ventral posterior temporal lobe, anterior cingulate cortex, right inferior lateral cerebellum and left prefrontal cortex are activated. Emotional aspects of language such as tone of voice seem to preferentially engage the right hemisphere.

The domain of language, like those of vision and attention, are a collection of networks specialized for the execution of quite specific functions. The elucidation of these functions remains to be completed. The study of language clearly illustrates how the original localizationist models of brain function are being systematically expanded though the use of new brain imaging techniques. Just as the microscope revolutionized biology, and the telescope revolutionized astronomy, so brain imaging may revolutionize cognitive neuroscience and its associated models.

## 4. MEMORY

The models we have considered so far emphasize modularity and parallel distributed processing. The same principles apply to models in the domain of memory research, with recent research suggesting a large number of independent memory stores in the brain related to many different cognitive processes. Memory: it would appear, is another example of a horizontal process. There is no place in the brain that one can point to as the center for memory: there is no single activity that encompasses all that this term denotes.

Because memory systems occur in parallel, selective brain damage can harm one type of memory while leaving others intact. Witness for example patients with prosopagnosia, who show a selective loss of the ability to recognize faces but may recognize other objects without any problem. It seems that their 'face memory' and processing structures have been damaged while other memory stores remain intact. The case of prosopagnosia illustrates another interesting point that is implicit within the parallel distributed processing approach — there does not need to be a firm distinction between processing and memory structures in the brain. Both are likely to engage the same networks. The history of the processing structure, which is its memory of past processing and events, is entwined with its ability to process memory and both are encoded within the pattern of connections between neural computing units.

In research over the past 30 years, memory researchers have been particularly interested in patients with temporal lobe lesions and have found that the hippocampus and surrounding cortical areas play an essential role in the formation of new memories for conscious information, such as facts and events (Zola-Morgan and Squire, 1993). Psychologists often refer to this type of memory as 'declarative or explicit memory', because subjects are consciously attempting to store or retrieve remembered information. Bilateral damage to the hippocampus leads to severe deficits in declarative memory, where the patient cannot store new information about new events occurring in the world. Recall of old stored information however, can proceed independent of a fully functioning hippocampus.

In people with hippocampal damage, recall of memories formed before the illness or accident that damaged the brain shows only a graded impairment, with memories formed long before the illness or injury remaining intact. This shows that while temporal lobe structures are essential for new memory formation, this brain region is not the repository of existing memories. Where might long term memories be stored?

Not all types of memory are affected by hippocampal damage. People with hippocampal damage can still learn and remember new motor skills even through they are incapable of explicitly remembering practicing the task. This so-called 'procedural memory' for skills and habits may preferentially involve the basal ganglia and the cerebellum.

Hippocampus damaged patients can show very basic forms of associative learning such as classical conditioning, which involve structures such as the cerebellum and amygdala. This type of memory is referred to as 'implicit memory' because it is an unintentional, non-conscious form of retention and retrieval. Amnesics also show perceptual priming effects. Although amnesics may be unaware that they have viewed perceptual nonsense shapes in the past, they have shown that they have implicitly encoded and can recollect such shapes, by performing well above chance on recognition tasks involving these shapes. Such visual perceptual priming effects appear to involve the perceptual systems of the occipital cortex. So these spared memory functions following hippocampal damage strongly support the notion that there are numerous types of parallel memory systems distributed throughout the brain, acting in concert. Study of implicit memory has dramatically increased since the early 1980s.

Psychologists, via lesion studies and animal research, have also distinguished a type of memory called 'working memory' which allows task specific information to be temporarily accessible over short delays, like remembering a telephone number long enough to be able to dial it or write it down. The prefrontal cortex has been particularly associated with working memory, as well as with the time coding of memories. Researchers in human performance labs have identified that there are unique, independent temporary stores for visual and auditory working memory.

The work of the Canadian neurologist Penfield showed that stimulation of cortical regions, particularly the temporal lobe, could produce a vivid flashback of a particular experience in patients, suggesting the cortex as one reservoir of long term memories. Since that time there has been an emerging consensus that long term memories are widely distributed across the brain. Hebb's cell assembly theory as previously described, is the mechanism through which learning might take place at the neuronal level, and it provides a mechanism for association to occur. The increased strength of connections among synapses in a Hebbian network would form the basis of memory and predispose the network to be activated in future similar occasions. Since Hebb's time, experimental support for this model has been forthcoming. Researchers have found enhanced activation of neural activity with repeated stimulation, a process now known as long term potentiation (LTP).

While there are many types of memories, and many processes associated with memory including encoding, storage, retrieval and consolidation, as well as many scales of function, there are therefore rudimentary models of memory that can be systematically explored and extended.

## 5. EMOTION

While the early schools of psychoanalysis and behaviorism — which shunned brain-based analysis of behavior — provided deep conceptual analyses of the nature of emotion, modern neuroscience has trodden cautiously but emphatically into this area. Emotion is generally thought to provide a 'value system' in the brain which allows the assessment of the significance of current percepts in relation to past memories, and the selection of appropriate behaviors according to current motivational state. Also intimately tied to emotion are bodily changes mediated by the autonomic nervous system — for example palpitations, hyperventilation and sweaty palms of fear.

In modeling the neural bases of emotion, a key network has been the limbic system. As seen in the chapter on the brain's anatomy, the limbic system is an amalgamation of subcortical and allocortical regions including the amygdala, septal area, hippocampus, hypothalamus, mamillary bodies and adjacent regions of cortex — particularly the orbitofrontal cortex, cingulate gyrus and parahippocampal gyrus. Many of these brain regions are evolutionarily primitive and are thought to be involved in the control of basic bodily processes and primitive behaviors.

The role of the limbic system in emotion is suggested in studies of patients with temporal lobe epilepsy, whose seizures can lead to abnormal activation of limbic structures including the amygdala and hippocampus. In some cases, these patients experience aggression, fear, sexual arousal, uncontrollable laughter and even bliss during their epileptic seizures. In addition, electrical stimulation of the amygdala or hypothalamus can induce aggression or defensive behavior, while stimulation of the septum can enhance sexual responses. Stimulation of certain parts of the hypothalamus can induce eating and drinking.

Despite these associations between emotions and specific brain sites, we still do not really know at a brain function level what emotions are. Furthermore, the limbic system model of emotion is a very general one. More recently, neuroscientists have focused on individual circuits within the limbic system and their relation with specific aspects of emotions, such as fear, anger, pleasure, disgust and pain.

For example, the American behavioral neuroscientist Joseph LeDoux (LeDoux, 1995) has identified a network centered around the amygdala that he proposes plays a major role in fear and anxiety in humans and other animals. The amygdala is particularly well placed for evaluating the emotional value of current sensory input according to past memories. It receives rich sensory input from the thalamus and cortex, receives memory input from the hippocampus, sends autonomic output to the body via the hypothalamus and brainstem, and can direct behavioral output via the periaqueductal grey. Le Doux has shown that destruction of the amygdala leads to animals failing to learn the emotional significance of stimuli and showing decreased autonomic arousal and fear-like behavior to fear-inducing stimuli. In humans, amygdala damage can lead to aberrant sexual behavior, disinhibited social behavior and physical aggression.

While the amygdala seems important for negative emotions such as fear, other circuits in the limbic system seem to mediate positive emotions such as pleasure. Many human beings stimulate their own brains for pleasurable effects by taking drugs. The neural substrates underlying drug reward have been an area of huge research interest over the past decade, with a large body of literature, principally from animal research, pointing to the importance of the neurotransmitter dopamine and a circuit involving the ventral tegmental area, the nucleus accumbens and the prefrontal cortex. Drugs of abuse such as nicotine, alcohol, heroin, cocaine and amphetamine all appear to selectively activate this dopamine pathway in the brain. Activation of this pathway is also involved in the pleasurable effects of natural rewards such as food, water and sex (Koob, 1992).

As proposed in The Big Picture (bringing together Le Doux, Damasio and Halgren's biological models of emotion), biological models of emotion are providing an increasingly detailed platform, from which to link to psychological and pschyodynamic models of emotion. Models of emotion demonstrate that even our highest levels of conceptualization are considered as plausible phenomena to be understood at least to some extent, and explicitly modeled.

## CONCLUSION

This chapter has presented selected models of perception, attention, language, memory and emotion that are emerging from psychology and its scientific partners. In recent years there have been substantial advances in our understanding of the mind, although we clearly have a long way to go with regard to understanding the detailed mechanisms.

One striking aspect of the models described is that very few could be claimed to be the exclusive province of psychology. The models we have considered have emerged through a multidisciplinary effort involving psychologists working with neurologists, anatomists, pharmacologists or computer scientists. One gets the overwhelming impression that the future of modeling the mind lies in multidisciplinary collaboration and in complimentary diversification rather than specialization. The study of cognition is no longer the exclusive province of psychology as it was for many decades. Multidisciplinarity has changed and will continue to change the landscape of psychology, and this 'Integrative' symbiosis seems likely to yield increasingly realistic models of human brain function and behavior.

## REFERENCES AND SUGGESTED FURTHER READING

Barlow, H. B. (1972) Single units and sensation: a neuron doctrine for perceptual psychology? *Perception,* **14**, 371–94.
Biederman, I. (1987) Recognition-by-components: a theory of human image understanding. *Psychological*

*Review,* **94**(2), 115–47.

Bridgeman, B., Kirch, M., and Sperling, A. (1981) Segregation of cognitive and motor aspects of visual function using induced motion. *Perception and Psychophysics,* **29**, 336–342.

Crick, F. H. C. (1994) *The astonishing hypothesis: The scientific search for the soul.* New York: Charles Scribner's Sons.

Damasio AR (1994) *Descartes Error.* Picador

Dennett, D. C. (1992) *Consciousness explained.* Boston: Little, Brown and Co.

Drevets, W. C., Burton, H., Videen, T. O., Snyder, A. Z., Simpson, J. R., Jr., & Raichle, M. E. (1995) Blood flow changes in human somatosensory cortex during anticipated stimulation. *Nature,* **373** (6511), 249–52.

Edelman, G. M. (1992) *Bright air, brilliant fire: On the matter of the mind.* New York, BasicBooks, Inc.

Fodor, J. A. (1983) *Modularity of mind.* Cambridge: MIT Press.

Gazzaniga, M. S. (1995) *Consciousness and the cerebral hemispheres.* In M. S. Gazzaniga (Ed.), The cognitive neurosciences, Cambridge, MA, MIT Press.

Goltz, F. L. (1888) Uber die verrichtungen des grosshirns, English transl. in. In G. v. Bonin (Ed.), *Some Papers on the Cerebral Cortex,* (pp. 118-158). Springfield: Charles C. Thomas, 1960.

Harrington, A. (1991) Beyond phrenology: Localization theory in the modern era. In P. Corsi (Ed.), *The enchanted loom:* Chapters in the history of neuroscience. History of neuroscience., (Vol. 4, pp. 207–239). New York, Oxford University Press.

Jacobsen, C. F. (1932) Influence of motor and premotor area lesions upon the retention of killed movement in monkeys and chimpanzees. *Proceedings of the Association for Research into Nervous and Mental Disorders,* **13**, 225–247.

Koob, G. F. (1992) Drugs of abuse: anatomy, pharmacology and function of reward pathways. *Trends in Pharmacological Sciences,* **13**(5), 177–84.

LeDoux, J. E. (1995) Emotion: clues from the brain. *Annual Review of Psychology,* **46**, 209–35.

Livingstone, M. S., and Hubel, D. H. (1988) Segregation of form, color, movement, and depth: anatomy, physiology, and perception. *Science,* **240**, 740–749.

Mesulam, M. M. (1990) Large-scale neurocognitive networks and distributed processing for attention, language, and memory. *Annals of Neurology,* **28**(5), 597–613.

Milner, A. D., and Goodale, M. A. (1995) The visual brain in action: *Oxford University Press;* Oxford, England.

Mishkin, M., Ungerleider, L. G., and Macko, K. A. (1983) Object vision and spatial vision: Two cortical pathways. *Trends in Neurosciences,* **6**, 414–417.

Piattelli-Palmarini, M. (1989.) Evolution, selection, and cognition: From 'learning' to parameter setting in biology and in the study of language. *Cognition,* **31**, 1–44.

Posner, M. I., and Dehaene, S. (1994) Attentional networks. *Trends in Neurosciences,* **17**(2), 75–9.

Posner, M. I., and Petersen, S. E. (1990) The attention system of the human brain. *Annual Review of Neuroscience,* **13**, 25–42.

Raichle, M. E. (1994) Images of the mind: studies with modern imaging techniques. *Annual Review of Psychology,* **45**, 333–56.

Rumelhart, D. E. (1989) The architecture of mind: A connectionist approach. In M. I. Posner (Ed.), *Foundations of cognitive science,* (pp. 133–159). Cambridge, MA, MIT Press.

Schneider, G. E. (1969) Two visual systems. *Science,* **163**(870), 895–902.

Singer, W., and Gray, C. M. (1995) Visual feature integration and the temporal correlation hypothesis. *Annual Review of Neuroscience,* **18**, 555–86.

van der Heijden, A. H. C. (1992) *Selective Attention in Vision.* London, Routledge.

Zola-Morgan, S., and Squire, L. R. (1993) Neuroanatomy of memory. *Annual Review of Neuroscience,* **16**, 547–63.

# Chapter 13

# MODELS OF THE BRAIN IN PSYCHIATRY

Graeme C Smith

Department of Psychological Medicine, Monash Medical Centre

Three defining characteristics of psychiatry will shape this chapter. First is the fact that psychiatry deals with disorders of emotional life. So too does psychology, with which it shares techniques and should therefore share models of the brain. Second, psychiatry is a branch of medicine, in particular it uses the methods and shares the models of neurology. Third, psychiatry emphasizes the fact that the patient is part of a social matrix, and a social model of the brain would be relevant. In practice, psychiatrists use these three types of model in parallel, in a systems theory *biopsychosocial model*. There should therefore be a biopsychosocial model of the brain; this chapter examines the extent to which that is true, drawing on past and present mainstream models. We should note at this stage that there is a paucity of integrated models that would truly support a biopsychosocial approach.

The biopsychosocial model is immediately problematic (Smith, 1991). It has to reconcile the disparate contributing models described above, which in turn embrace both biological and psychological ways of thinking. Psychiatrists operate in their predominantly biological mode when they believe that mental phenomena are products of brain activity, when they examine a patient and look for a physical cause of a mental disorder, and when they use physical treatments such as drugs and electroconvulsive therapy. They operate in primarily psychological mode when they believe that mental phenomena function independently of physical brain processes, when they take the personal history of the patient, and when they examine the psychological system — the mental state examination — inquiring into the patient's feelings and perceptual experiences. A psychological mode of operation is most evident when they engage in psychotherapy helping patients to understand their patterns of behavior, the meaning of their symptoms and to develop ways of dealing with them psychologically and socially (see next chapter on psychotherapy).

# THE IMPORTANCE OF MODELS OF EMOTION TO PSYCHIATRY

In developing biopsychosocial models, we must address the fact that clinical psychiatry is based on the construct of mental disorders, clusters of phenomena which seem to have a consistency of presentation that allows them to be considered as categories of disorder, and perhaps a diagnosis. Psychiatrists recognize that this reductionist activity does not do full justice to the complexity of human mental functioning, and are at pains to consider the patient as an individual, not as a disorder. However, categorization has considerable value in predicting choice of therapy and outcome, and for communication.

To the extent that the phenomena constituting these categories are measurable, they form the basis of research. Modern psychiatric research tends to use continuous rather than categorical measures. For construction of animal models, those phenomena which are common to humans and animals are used. The phenomena on which diagnostic categories are based are symptoms reflecting disturbed attention, emotion, perception, thinking and behavior. Of these, emotion is the one which is central to psychiatry, yet is the most difficult to define and to identify in animal models. Lets briefly explore various aspects of emotion (which is also explored in The Big Picture and psychology chapters), before moving onto models of anxiety disorders, depression and schizophrenia.

It is generally agreed that emotion is a complex feeling state with mental, somatic (body) and behavioral components (LeDoux, 1987). The mental component involves experience, evaluation and organization of behavior, and this can be reflexive or voluntary, conscious or unconscious (Ohman and Birbaumer, 1993). It is a feeling state that cannot be conjured up. It is as if it has been placed there by someone else, akin to the sense that there is another 'I' which is me, which controls my behavior but over which I seem to have no control. It is something that belongs to us because of what we are, rather than what we do (Zizek, 1991). Neurotic patients seek the help of mental health professionals in coming to terms with this situation and gaining some sense of control. Psychotic patients project the 'I' onto external agents and want them controlled.

Definition of the requisite stimulus context for emotional experience and behavior is required. The *meaning* of a stimulus, rather than the stimulus itself, is what determines whether or not emotion is elicited. The mark of the occurence of an emotional stimulus is the emergence of behavior which has neither sufficient not adequate external purpose or reason. The terms 'inefficiency' or 'irrationality' are often used when such behavior is observed. There may be unusual intensity or emphasis, or a sudden absence of behavior. The explanation is sought within the organism rather than within the stimulus.

To a large extent most of the current work on animal models of emotional behavior assumes that it is equivalent to the model of learned, potentially rewarding behavior. In the case of humans, it is argued that emotional phenomena are reflected in noninstrumental behavior, that is, behavior which is not shaped by its consequences (Ohman and Birbaumer, 1993). This argument is used to explain the irrationality or excess that characterizes emotional behavior. This apparent conflict may be a semantic problem, since common to both domains is the concept that emotional behavior is not maintained by its consequences but rather by the internal motivation to achieve a reward.

In this sense, all psychological activity (as opposed to reflex activity) takes place in an emotional context, since we are always either avoiding or approaching something, and this is the major dimension on which significance operates. Arousal and to a lesser extent dominance are the other major dimensions (Ohman and Birbaumer, 1993).

## Dimensions of emotion

Common language would suggest that there are fundamental and distinct emotions, easily identifiable by those experiencing them, and each accompanied by distinct behavioral outputs. Darwin suggested that on a phylogenetic basis, fundamental emotions occur across species that they are genetically determined. A similar argument has been proposed for temperament (Cloninger *et al.*, 1993). Darwin's argument depended on the validity of facial expression as a marker of emotions. This topic has received much attention in recent years, but remains controversial (Izard, 1994). An alternative explanation is that facial expressions represent responses to specific appraisals of stimuli rather than to the emotional content.

Because of this and other concerns about the validity of the concept of fundamental emotions, there has been increasing emphasis on the dimensional concept (Ohman and Birbaumer, 1993). The major dimensions are valence and arousal (Lang *et al.*, 1993). They define the general direction of behavior (appetitive or defensive) and the amount of energy resources to be devoted, without specifying the patterns of response. Emotions are subordinate to this 'strategic state' and tactical with relation to it: fear, aggression and pleasure are three broad categories, and the finer divisions of emotional responses may be differentially recruited under these headings (hence the difficulty of classifying them).

## Theories of emotion

Feeling states cannot be studied directly. We have to resort to studying those related emotional phenomena which are measurable, such as verbal reports, somatic changes and behavior — as they occur in stimulus contexts defined by us as emotional. Naming these domains of observation locates the feeling state of emotion within a network of functions involving sensory perception, attention, association, cognition, memory, motivation and learning, together with organization of behavior and motor, verbal, visceral and endocrine responses.

Important though it is in the common language and in psychiatry, the phenomenon of conscious experience of emotional feeling is not the focus of all theories. For some it is an epiphenomenon, the labeling process. Most neurobiological work assumes that it is not necessary to take it into consideration in the development of models.

Is the subjective experience of emotion a *sine qua non* of the occurrence of an emotional event in humans? One problem with accepting this thesis is that subjective experience is not directly observable. We depend on verbal reports for evidence of its occurrence. Another problem is that there appears to be a group of people (labeled alexithymic) for whom, either as a trait (inherent) or state (transient), subjective experience of emotion is absent, yet their behavior and bodily responses betray the presence of emotion. A third problem is that if we require the subjective experience of emotion, we remove the possibility of using animal experiments to explore the topic. However, there is abundant evidence for the relative independence of the cognitive, affective, somatic and behavioral aspects of emotion (Lang, 1993); this helps to validate the use of animal experiments in the development of models of the brain for emotion.

Thus we arrive at a model of emotion which holds that it is a set of mental, neuro-biological, somatic and behavioral events, which occur in a stimulus context which has significance for the individual because of past experience. This permits us to dissect out and describe the components. The emphasis on significant stimuli takes us immediately to the theories about how the mental apparatus is constructed to evaluate the meaning of stimuli. The relevant theories are psychoanalytic, cognitive, behavioral and social.

The psychoanalytic theory of emotion has to do with the formation of the subject (Zizek, 1991). It holds that this is based on instinctual drives, but is determined definitively by experience, that is, by what is made of what is learned. Thus the subject unconsciously develops objects of desire, and ways of dealing with the consequences of this. Whilst these processes shape the type of emotion that will dominate that particular subject's experience and moderate its intensity, they also produce the behavioral and somatic symptoms which are the visible hallmark of the unconscious processes which have occurred. The subjective experience of emotion is not a requisite of this theory.

Cognitive and behavioral theories have become somewhat blended. According to the model developed by Lang and his colleagues (Ohman and Birbaumer, 1993), emotion is represented in memory as interconnected nodes comprising stimulus response and meaning information, just as for any other information. If inputs match the information held, an emotional response ensues; the intensity of input needed will depend on the associative strength within the emotional network. The network includes one component that analyses the stimulus in a general sensory way, a second that analyses meaning, and a third that holds memory of the pattern of somatic and verbal responses. Variations in the way that this network is organized are held to account for the varying susceptibility to emotional activation between individuals, and to the variations in the favored form of expression, for example, panic attacks, phobias or generalized anxiety (Cuthbert and Melamed, 1993). A component for subjective experience is not required in this model, but within the cognitive/behavioral literature there is debate about this (Cuthbert and Melamed, 1993). Lang now defines emotion as an action set, defined by a specific information structure in memory, which when accessed is processed as both a conceptual and a motor program (Lang, 1993).

Social theories of emotion are summarized by Kemper (1987). One such view is that emotions are multireferential lay constructs, that is, ones that are in everyday common language use and which have multiple public and private referents. They allow individuals to organize, label and talk about certain perceptual experiences.

None of these theories are mutually exclusive. Conditioning of emotional responses does not require conscious awareness (Ohman, 1988). Cognitive theorists accept the importance of developmental phases in vulnerability (Ohman, 1986). Experimental acquisition of emotional responses is only partially explained by conditioning (Sartory, 1993). Concepts invoked to explain this include those of genetically prepotent stimuli, preparedness and pre-attentiveness (Ohman, 1993). In the psychoanalytic perspective, becoming a subject means experiencing oneself as an object, a victim in which we assume that which is imposed upon us.

Other models of emotion fall into line with the psychoanalytic understanding. In a sense they operationalize it, reducing it to components that can be studied by use of the scientific method, and influenced by cognitive, behavioral and pharmacological means.

## Neural basis of emotion

In the 1940s Papez and MacLean popularized the notion of a limbic circuit as the basis of emotional life. Their model has not withstood empirical investigation, though revisions of some components of it constitute the current models. There would be advantages in confining the use of the term limbic to its original, anatomical one. However, it has come to have a functional meaning, implying emotional processes, and that is how we will use it here.

Extensive experimental work in animals has produced a coherent picture of how the brain is organized to deal with emotional phenomena. Whereas in Papez' model of the limbic circuit, the hypothalamus and cingulate cortex were given special prominence, today the focus is on the amygdala and the ventral striatum, two structures not included in Papez' circuit (see details of limbic system in the anatomy chapter). The major paradigm used is that of the study of events that occur in the context of a stimulus that the animal has learnt is potentially rewarding (Scheel-Kruger and Willner, 1991). In essence, the model holds that stimuli are assessed for their rewarding capacity in the hippocampal system, fixed for attention in the amygdala system, and in the ventral striatum occurs the final process of integration and evaluation and selection of appropriate somatic and behavioral responses. The mesolimbic dopaminergic neuronal system is essential to this process. (Also see the Damasio's model of emotion in The Big Picture and psychology chapters).

## BRAIN MODELS OF PSYCHOPATHOLOGY

Much of the data used to construct brain models in psychiatry comes from studies on patients with a psychiatric disorder. Nevertheless there is a body of knowledge from animal experiments which informs most of the current models. A growing concern about the diffuseness of brain models in psychiatry and the difficulty of testing them has led to a call to focus on hypotheses that are testable, and for a program of rigorous testing of such hypotheses (Carpenter et al., 1993). This book highlights the rich multidisciplinary landscape from which models of psychopathology seem likely to be further developed and tested.

## Anxiety disorders

In thinking about emotion, we run into the problem of deciding whether or not there is a continuity of the spectrum from normal state to mental disorder. It is a contentious issue. We will use anxiety and its disorder to illustrate this, and to introduce us to the modern approach to establishing models of the brain in psychiatry.

Anxiety is a normal emotion which probably functions to focus attention appropriately at times of threat and to organize appropriate somatic and behavioral responses (see the 'fight and flight axis' in The Big Picture). But it may be of such a degree and type that it distresses the patient and is disabling, especially if it is occurring when there is no threat apparent. It may then be experienced as a chronic feeling of tension, with vague somatic feelings such as palpitations, diarrhoea and headache. It may occur episodically, as panic attacks for no apparent reason. It may present as symptoms used to relieve anxiety — rituals and obsessive thoughts, as in Obsessive Compulsive Disorder. Much of the research on humans has been done on these latter categories.

Establishment of unique brain models for each diagnostic category in psychiatry involves a rigorous exploration of its component parts. This has not yet been done for anxiety. What we do have are some biological correlates of particular forms of anxiety disorder, such as Obsessive Compulsive Disorder. Patients with this disorder seem to have increased blood flow in the orbitofrontal cortex, neostriatum, globus pallidus and thalamus when they experience urges to perform compulsive movements, and in the hippocampus and posterior cingulate cortex when they experience anxiety (McGuire *et al.*, 1994). The anterior cingulate cortex has been implicated in other studies.

Observations such as these help pinpoint the neural systems likely to yield data about causality if further explored. They do not tell us this directly. For instance, other clinical research data point to a norepinephric (NE) oversensitivity in panic disorder (Nuts and Lawson, 1992). In order to understand what that might mean, we have to explore the function of the NE system under normal conditions.

In the anatomy chapter, the authors describe the extensive neuronal system that uses NE as its transmitter. Reference is made to the possible role of the locus coeruleus NE system in attention. In the brain dynamics chapter (10a), we see how important the NE system is in facilitating the intrinsic cortical dynamical activity, and modulating the pattern of cortical firing according to the aroused state.

The NE system, together with the other chemically defined brainstem systems, including the dopaminergic (DA), serotonic and acetylcholinergic (Ach) constitute the reticular system, first described in electrical terms by Moruzzi and Magoun in the 1940s. Behavioral studies in animals, including direct recording from the locus coeruleus in conscious monkeys, has confirmed that these NE neurons are most active just before the animal shows evidence of enhanced arousal or attentiveness (Jacobs, 1987). The end result is a facilitation of cortical sensory processing and enhancement of cognitive processes that guide behavior.

This evidence from behavioral studies in animals is consistent with the clinical studies. Attention is regarded as a key feature of anxiety. But what about the behavioral features? To some extent this has been explored: in monkeys, the behavioral phenomena that we associate with anxiety in humans can be mimicked by manipulation of the NE system (Redmond, 1987).

Unfortunately, the study of NE activity is complicated by the fact that it is distributed in the autonomic nervous system as well as the brain. However, we do know that drugs which produce a NE challenge can produce feelings of anxiety in normal subjects, and induce panic attacks in those prone to them.

NE is only part of the story. The serotonin system is also implicated in anxiety. A number of the drugs effective in ameliorating anxiety disorders are serotonin re-uptake inhibitors, and a serotonin agonist is anxiogenic in both normal subjects and those with anxiety disorders (Nutt and Lawson, 1992). The serotonin system modulates intrinsic cortical electrical activity, inhibiting the margins of neuronal pools. It modulates brain activity in general. Behavioral studies suggest that the serotonin system constrains information processing in a way that stabilizes the process and results in a controlled behavioral output (Spoont, 1992). Disruption of the serotonin system thus disturbs a whole range of behaviors, including sleeping, eating, sexual and aggressive behavior (Jacobs, 1987). It is thus easy to conjecture that the serotonin system is an important one for human anxiety states, but much harder to prove that it is.

The amino acid GABA is so closely implicated with anxiety, through its co-location with benzodiazepine receptors, that it must receive our attention (Nutt and Lawson, 1992). GABA is the neurotransmitter in some 40% of CNS neurons, mainly in local inhibitory circuits. However, there is considerable heterogeneity produced by the various receptor subtypes. Of particular relevance to psychiatry would seem to be the observation that GABA is the transmitter for the fronto-reticular system which imposes a patterned descending tone on the reticular formation.

Positron Emission Tomography (PET) studies on patients with Obsessive Compulsive Disorder implicate two of the basal ganglia-thalamocortical circuits. These are the orbitofrontal component of the prefrontal circuit and the limbic circuit (details in Alexander model in The Big Picture and anatomy chapters). Consideration of these will allow us to illustrate the way in which the original concept of the limbic system has been dismantled, and replaced by one based on empirical anatomical and behavioral studies.

The lateral orbitofrontal circuit rather than the dorsolateral circuit is the one implicated in Obsessive Compulsive Disorder (McGuire *et al.*, 1994). Disruption of it, in monkeys, produces impairment of the ability to make appropriate changes in behavioral set (Goldman Rakic, 1987).

The ventral striatum, which includes the nucleus accumbens, is a focal point in the newly defined limbic circuit. In this model, it is regarded as receiving information about potentially rewarded ('emotional') stimuli that has been assessed as such by the hippocampal system, and fixed for attention by the amygdala system (Scheel-Kruger and Willner, 1991). The limbic circuit's closed cortical loop involves the anterior cingulate and medial orbitofrontal cortex. Through its output to other brain systems, the autonomic, endocrine and motor behavioral components of emotional responses are initiated. Both the NE and serotonin systems play an important part in this process, through both their cortical and subcortical innervations. Disruption of this system produces disturbances in emotional behavior, as evidenced by changes in cognition and behavior. Further detailed dissection of this complex activity is required before we can relate it to specific emotional disturbances. Further work in humans is required to explore the feeling component.

## Depression

Keep these systems in mind as we now turn to another apparently discrete emotional disorder, that of mood. Depression can be defined in quite variable ways. The central, most consistent features defining depressive disorders are feelings of depression (blue, dispirited, feeling like giving up, a lack of motivation) together with a distinct drop in self-esteem or self-worth, with the individual being self-critical or self-reproachful (Parker *et al.*, 1994).

Many models of depression have been created. Amongst the non-biological ones are the psychoanalytic concept of inwardly directed anger, the cognitive 'wrong attribution'

model, the behavioral model of learned helplessness, and the social model of depresso-
genic stressors. All of these have strong intuitive support, and the psychological models
which claim to be based on such theories are often as effective as antidepressant drugs in
ameliorating mild and moderate forms of depression. But particularly for severe depression, where biological treatment is essential, there are biological signs which invite a biological model lending itself to rigorous hypothesis testing (Parker *et al.*, 1994).

In severe depression, also called melancholia, the quality of the depression is
somewhat characteristic; it is difficult to cheer the patient up, and there may be guilty
rumination. There may even be delusions of worthlessness, in which case the description 'psychotic' is applied. Biological or 'vegetative' signs are frequent and multiple:
disturbed sleep, lowered appetite, weight loss, loss of libido, impotence, constipation,
slow speech, slow thoughts, impaired concentration and slow movement. Sometimes
there is agitation. The patients can so resemble those with dementia or Parkinson's
Disease that they are misdiagnosed as having these conditions, and vice versa.

Recent brain imaging studies suggest that there may be dysfunction of prefrontal
cortex and caudate nucleus in patients with severe depression, though it is not clear
whether this is a state or trait effect (Austin and Mitchell, 1994). Other biological
models of severe depression have proposed involvement of diverse networks, including: hypothalamo-pituitary-adrenal dysregulation (manifested as failure to suppress
the diurnal cortisol rhythm); other circadian rhythm disturbances such as that of
sleep; electrical kindling phenomena; and various disturbances of norepinephrine,
serotonin, endorphins and second messenger G proteins (Campbell *et al.*, 1994).
Often these models have more to do with describing modes of action of antidepressant drugs than explicating the biology of depression. Nevertheless they have been
useful in the development of new drugs.

Thus several of the systems already discussed are implicated, and we are introduced
to the dopaminergic systems. Those that concern us here are the nigrostriatal and the
mesocortical systems. The nigrostriatal system is an important component of the 'motor'
circuit. It has complex modulatory effects on motor behavior, but overall it seems to
reinforce cortically initiated action. It could do this during the preparation for a motor
response, since it acts on both the direct and indirect components of the 'motor circuit'.

The mesocortical system shows a major evolutionary development in primates.
Whereas in the rat it innervates only the prefrontal cortex, in primates it innervates
the whole cerebral cortex, in keeping with the more widespread distribution of dorsomedial thalamic nucleus.

Thus it innervates the dorsolateral prefrontal cortex. This is the cortical component of the other prefrontal circuit, the dorsolateral circuit. Prefrontal cortex has the
ability to access and hold on line information relevant to the task at hand, even
though the information is not contained within the current stimulus (Goldman-
Rakic, 1987). The mesocortical dopaminergic system seems to be involved in the
preparatory coping behavior that is elicited by a cognitive challenge or other stressor
(Scheel-Kruger and Willner, 1991). It is particularly sensitive to stressors. It modulates the electrical activity of the cortex.

A recent review of brain imaging findings in depression, and models of kindling
appropriate to this disorder, is found in George *et al.* and Post *et al.* (1998).

## Schizophrenia

The third major psychiatric disorder to consider is Schizophrenia. The characteristics of patients given this diagnosis include disturbances in the external expression of emotion (affect), disturbances in thinking, and abnormal perceptual experiences. Thus they may present with thoughts that indicate poor reality testing (delusions), and hallucinations (distorted sensory experiences in the absence of appropriate stimuli).

The most studied of all the psychiatric disorders with respect to brain models, schizophrenia remains an enigma. This is partly to do with problems of classification, but also with premature claims for the specificity of findings. So extensive has been the research that we have available a number of fine general reviews for example (Pilowsky, 1992; Corcoran and Frith, 1993), as well as critical reviews of neuropathological findings (Benes, 1993), the dopamine hypothesis (Davis, 1991), and electrophysiology (Tueting, 1991), and novel theories concerning development (Crow, 1993), sleep (Home, 1993) and viral infection (Butler and Stieglitz, 1993).

The dopamine (DA) theory of schizophrenia remains the strongest one and serves as an orientating model. It was developed on the basis of the observation that drugs effective in ameliorating the symptoms of schizophrenia block DA receptors, and that their ability to do so correlates with their clinical potency. The theory was strengthened by post-mortem studies which appeared to demonstrate an increased density of several DA receptors in the striatum of patients with schizophrenia. As re-formulated by Davis *et al.* (1991), the DA hypothesis states that schizophrenia can be characterized by *hypo*dopaminergia in mesocortical and *hyper*dopaminergia in mesolimbic DA neurons in some patients, and that the two conditions may be related. There is nothing to suggest that these altered DA states are related to etiology or pathogenesis, and it is unlikely that they are universal features or sole disturbances. They may well be related to only some symptoms in some patients, and they may not be specific for schizophrenia nor indeed for disorder. Nevertheless they constitute a robust finding.

This introduces us to the mesolimbic component of the DA systems. Its major target is the nucleus accumbens of the ventral striatum, and through it comes to have a major modulating influence on the 'limbic' circuit. It plays a key role in translating motivation into action, though not for instinctual drives, rather for those rewards that have been learned (Scheel-Kruger and Willner, 1991). Its activation is necessary if behavior likely to produce such rewards is to be selected, and its continued activity is required if the goal is to be obtained.

It will be recalled from earlier in the chapter that the current model holds that it is in the hippocampal system that potentially rewarding stimuli are assessed, in the amygdala system that they are fixed for attention, and in the ventral striatum that the final process of integration of the 'limbic' circuit evaluation and selection of appropriate behavioral response patterns occurs. The mesolimbic system acts on all of these subcortical components (see figure 13.1).

There are many other influences as well, but this diagram conveys a view of how the brain is organized to deal with stimuli that have the potential to provide learned rewards, and which may be associated with feelings embraced by the terms emotion and mood.

The selection (above) of models pertaining to schizophrenia is offered only as a preliminary frame of reference. In this most complex of disorders there has been little

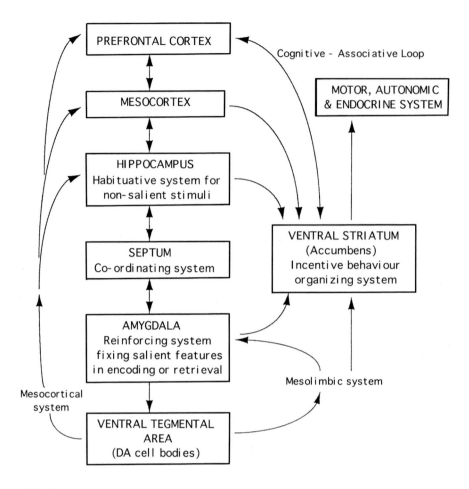

**Figure 13.1.** Network Interactions.

consensus as to the core disturbances in this disorder, but many postulates (for example, Gray *et al.,* 1991) suggest that hippocampal disturbances in mismatching input with stored information underlies the 'core' misattributions in schizophrenia, with subsequent adaptations (over-processing and then later shutting-down of information processing) resulting in predominant positive followed by negative symptoms respectively. Even the dopamine neurotransmitter dysfunction model has been challenged. For example, Arvid Carlsson, the father of this increased dopamine hypothesis, is currently trialling medications that do not block dopamine at all, but instead increase excitatory glutamate, since he now reasons that the problem might be associated with a glutamate versus dopamine *imbalance.* Elsewhere, there are multiple versions of new atypical antipsychotic medications that explore combinations of ratios of Dopamine 1/Dopamine 2 versus serotonin blockade. Classifications of this heterogeneous disorder, based upon patients' symptom profiles (for example, into Disorganization, Reality Distortion and Psychomotor Poverty Syndromes; Liddle, 1987; Williams, 1996) seem likely to make model development and testing in this disorder more tractable.

# CONCLUSION

It is not surprising that imaging, biochemical and pharmacological studies of patients with psychiatric disorders tend to yield results suggesting abnormal structure and function of these systems. Considerably more work is required in order to test the myriad of hypotheses and models that arise from the studies so far. In fact it would be true to say that rigorous model development and testing has barely begun in this field. We do not yet know why effective biological treatments such as electroconvulsive therapy and lithium work. And just as we think we are on our way to finding out, neuroscientists discover new neuronal circuits and new transmitters which have to be fed into the equation. Acetyl choline, epinephrine, histamine, nitric oxide and the plethora of peptides are just some examples. What is encouraging given the inadequate fundamental understanding of these disorders, is that for so many people who present with disturbances of their psychological system, therapies have been effective.

Psychological therapies alone are often not sufficient to ameliorate the painful symptoms of the psychiatric disorders we have considered in this chapter. Social factors must be attended to, and all psychiatrists would use physical biological treatments as well in severe cases.

Rational development of better physical treatments would be greatly aided by better brain models. In that the currently used models provide for psychological and social factors, they are relevant to the biopsychosocial model of clinical practice. However, the gap between theory and practice is enormous, indicative of the fledgling status of the study of models of the brain psychiatry. With the escalation of multidisciplinary involvement in this field, model driven advances in neuroscience hold considerable promise to elucidate the psychopathology in disorders that are perplexingly common in humans.

# REFERENCES AND SUGGESTED FURTHER READING

Austin M. P, Mitchell P. (1995) The anatomy of melancholia: does frontal-subcortical pathophysiology underpin its psychomotor and cognitive manifestations? *Psychological Medicine.*

Benes F. M. (1993) Neurobiological investigations in cingulate cortex of schizophrenic brain. *Schizophrenia Bulletin,* **19,** 537–549.

Butler L D, Stieglitz T. L. (1986) Contagion in schizophrenia: A critique of Crow and Done. *Schizophrenia Bulletin,* **19,** 449–454.

Campbell I. C, Marsden K, Powell J. F. (1994) Medicine Shifting depression. *Psychological Medicine,* **24,** 271–273.

Carpenter W. T, Buchanan R. W, Kirkpatrick B, Tamminga C, Wood F. (1993) Strong inference, theory testing, and the neuroanatomy of schizophrenia **50,** 825–831.

Cloninger C. R, Svrakic D. M, Przybeck T. R. (1993) A psychobiological model of temperament and character. *Archives of General Psychiatry,* **50,** 975–990.

Corcoran R, Frith C. D. (1993) Neuropsychology and neurophysiology in schizophrenia. *Current Opinion in Psychiatry,* **6,** 74–79.

Cuthbert B. N, Melamed B. G. (1993) *Anxiety and clinical psychophysiology: three decades of research on three response systems in three anxiety disorders.* In Birbaumer N and Ohman A (eds.) op. cit. 93 – 109.

Davis K. L, Kahn R. S, Ko G, Davidson M. (1992) Dopamine in schizophrenia: a review and reconceptualization. *American Journal of Psychiatry,* **148,** 1474–1486.

George M. S, Post R. M, Ketter T. A, Kimbell T. A, Speer A. M. (1998) *Neural mechanisms of mood disorder.*

Goldman-Rakic P. S. (1987) Circuitry of primate prefrontal cortex and regulation of behaviour by representational memory. In Geiger SR (ed). Handbook of Physiology, Section 1, Volume 5, *Higher Functions of the Brain,* Part 1. Bethesda: American Physiological Society.

Gray J. A, Feldon J, Rawlin J. N. P, Hemsley D. R and Smith A. D. (1991) *Behavioral and Brain Sciences,* 14, 1–84.

Horne J. A. (1993) Human sleep, sleep loss and behaviour. Implications for the prefrontal cortex and behavior. *British Journal of Psychiatry,* **162,** 413–419.

Izard CE. (1994) Innate and universal facial expressions: evidence from developmental and crosscultural research. *Psychological Bulletin,* **115,** 288–299.

Jacobs B. L. (1987) Central monoaminergic neurons: Single unit studies in behaving animals. In Meltzer HY (ed). *Psychopharmacology*: The Third Generation of Progress. New York: Raven Press: f159–169.

Kemper T.D. (1987) A Manichean approach to the social construction of emotions. *Cognition and Emotion,* **1,** 353–665.

Lang PD. (1993) *From emotional imagery to the organization of emotion in memory.* In Birbaumer N and Ohman A (op. cit.) 69–92.

Lang P. J, Bradley M. M, Cuthbert B. N. (1993) *Emotion, arousal, valence and the startle reflex.* In Birbaumer N and Ohman A (eds.) op. cit. 243–251.

LeDoux J.E. Emotion. (1987) In Geiger SR (ed). Handbook of Physiology, Section 1, Volume V, *Higher Functions of the Brain*, part 1. Bethesda: American Physiological Society, chapter **10,** 419–459.

Liddle P. F (1987) The symptoms of chronic schizophrenia. A re-examination of the positive-negative dichotomy. *British Journal of Psychiatry,* **151,** 145–151.

McGuire PK, Bench CJ, Frith CD, Marks IM, Frackowiak RSJ, Dolan RJ. (1994) Functional anatomy of obsessive-compulsive phenomena. *British Journal of Psychiatry,* **164,** 459–468.

Nutt D, Lawson C. Panic attacks. (1992) A Neurochemical overview of models and mechanisms. *British Journal of Psychiatry,* **160,** 65–178.

Ohman A. (1988) Nonconscious control of autonomic responses: a role for Pavlovian conditioning? *Biological Psychology,* **27,** 113–135.23.

Ohman A. (1986) Face the beast and fear the ace: animal and social fears as prototypes for evolutionary analyses of emotion. *Psychophysiology,* **23,** 123–145.

Ohman A. (1993) *Stimulus prepotency and fear learning*: data and theory. In Birbaumer N and Ohman A. op. cit. 218–239.

Ohman A, Birbaumer N. (1993) Psychophysiological and cognitive-clinical perspectives on emotion: introduction and overview. In Biorbaumer N and Ohman A (eds). *The Structure of Emotion.* Seattle; Hogrefe and Huber Publishers 3–17.

Parker G, Hadzi-Pavlovic D, Wilhelm K, Hickie I, Brodaty H, Boyce P, Mitchell P, Eyers K. (1994) Defining melancholia: Properties of a sign-based measure. *British Journal of Psychiatry,* 164, 316–326.

Pilowsky LS. (1992) Understanding schizophrenia; structural and functional abnormalities of the brain are present in the condition. *British Medical Journal,* **305,** 327–328.

Post RM, Weiss SRB, Ketter TA, Denicoff KD, George MS, Frye MA, Smith MA, Leverich GS (1998) *The Kindling Model: Implications for the Etiology and Treatment of Mood Disorders.*

Redmond DE Jnr. (1987) Studies on the nucleus locus coeruleus in monkeys and hypotheses for neuropsychopharmacology. In Meltzer HY (ed), Psychopharmacology: *The Third Generation of Progress.* New York: Raven Press 967–975.

Sartory G. (1993) *The associative network of fear: how does it come about?* In Birbaumer N and Ohman A (eds) op. cit. 193–204.

Scheel-Kruger J, Willner P. (1991) The mesolimbic system: principles of operation. In Willner P and Scheel-Kruger J (eds.). *The mesolimbic DA system: from motivation to action.* Chichester: *John Wiley and Sons* (pp 559 – 557).

Smith GC. (1991) The brain and higher mental function. *Australian and New Zealand Journal of Psychiatry,* **25,** 215–230.

Spoont MR. (1992) Modulatory role of serotonin in neural information processing: implications for human psychopathology. *Psychological Bulletin,* **112,** 330–350.

Tueting P. A. (1991) Electrophysiology of schizophrenia: EEG and ERPs. *Current Opinion in Psychiatry,* **4,** 7–11.

Williams L. M (1996) Cognitive inhibition and schizophrenic symptom subgroups. *Schizophrenia Bulletin,* **22,** 139–151.

Zizek S. (1991) *Looking Awry. An Introduction to Jacques Lacan Through Popular Culture.* Cambridge, MIT Press.

# Chapter 14

# MIND–BRAIN IN PSYCHOTHERAPY

Russell Meares

Department of Psychological Medicine,
University of Sydney and Westmead Hospital

There are two principal models of mind in western psychotherapy. They are clearly of an abstract kind. One focuses on 'ego' and the other on 'self'. They are complementary rather than alternative. Ego theory has one main author — Sigmund Freud. Self theory, on the other hand, is multi-authored, still evolving, and cannot be given the same finite status as ego theory. Carl Gustav Jung and Heinz Kohut have been the main clinical contributors to this tradition. William James is the principal descriptor of self.

Between these two systems is a third one, termed 'object relations', which, in historical terms, offers a bridge between the old theory of ego and the newer one of self.

Both Freud and James brought to the study of mind and psychology, perspectives which were new and beyond the tradition of philosophical assertion which preceded them. In essence, they were representatives of the emergent world of modern science. Freud was a neurologist. His thinking was influenced by evolutionary theory and by neurophysiology. James was originally an academic physiologist. He pioneered a new approach to psychology, a change from a mental philosophy to a physiological psychology. In 1875, he founded, at Harvard, the world's first psychological laboratory. A model of mind, which is adequate from the scientific point of view, must be testable. Neither Freud nor James developed their theories in an era in which this notion was current. Falsifiable hypotheses based on ego psychology are particularly difficult to frame, perhaps impossible, since the 'unconscious' is one of the core elements of this theory. Superficially, self psychology appears even more problematic. However, hypotheses can be developed using the manifestations of self (for example, one means of doing so involves a study of linguistic patterns and this subject will be approached later in the chapter). A summary of models focusing on ego, object relations and self is presented below.

## EGO

Towards the end of the nineteenth century, an intellectual climate arose in which mental life became a respectable object of scientific scrutiny. A major figure in this field of inquiry was the great neurologist Jean Martin-Charcot, who began to study hysteria and hypnosis in his work at Salpetriere Hospital, in Paris, where he had become a chief physician in 1862. The work going on at Salpetriere attracted students from all parts of the world. In 1885, Sigmund Freud was one of them. Freud made his understanding of hysteria the foundation stone on which psychoanalytic theory rested.

It became evident at Salpetriere that those patients who were at that time given the diagnostic label of hysteria suffered from some symptoms which had a basis in experiences of which they did not seem to be aware. One part of the mind seemed to be keeping away information from another part of the mind. Certain experiences did not emerge into the domain of consciousness. They concerned traumatic experiences and determined certain symptoms. Relief of at least some of the symptom came about through bringing the hidden material into the light of consciousness.

Although hysteria is a relatively uncommon disorder and not all of its symptoms can be explained in terms of the above model developed at Salpetriere, it nevertheless seemed to offer an approach, at least as Freud saw it, to the understanding of a wider group of disorders which at that time were called the neuroses.

The first model of brain-mind in a psychotherapeutic sense was a simple one consisting of conscious and unconscious domains, the mechanism of repression maintaining the division between them. Freud conceived the systems unconscious and consciousness each as having a different place ('topos') so that the model was called 'topographical'.

Although Freud's model seems at first sight to differ little from the ideas already expressed at Salpetriere, there was, underlying his preliminary model, a quite different system of thinking. In a manuscript written during 1895 but not published during his lifetime, he made it clear that he had embarked on an ambitious project which was truly his own. He wanted to establish a neurophysiology of mind. His basic postulate was bold, innovative, and still retains a certain intellectual excitement. His approach was to consider each single neuron as 'a model of the nervous system as whole' (Freud, 1887–1902, pp.359). This principle, which depends upon the assumption that the fundamental laws which determine the function of the basic unit of the brain will also operate on a whole brain scale, is a plausible one. The approach is reminiscent of Sherrington's who, upon the basis of the spinal reflex, built up the notion of a unified action of the nervous system depending upon integrated excitatory and inhibitory systems.

Freud's first postulate was 'the principle of neuronic inertia' (pp.356). He conceived the electrical charges which could be recorded in nerves as the manifestation of 'quantities of excitation in flow' (pp.356). This principle asserts that 'neurons tend to divest themself of quantity' (pp.356). Put another way, the organism, when stimulated, attempts to get rid of that excitation. A state of high excitation is associated with unpleasure. The impulse is always towards pleasure, towards the reduction of excitation. However, as Pribram explains Freud's theory, 'transmitted excitations are not the whole story. Excitation can build up within a neuron and this increase need not necessarily lead to conducted impulsive activity' (Pribram, pp.402). Freud suggested that a neuron may become 'filled' — that is *cathected* with excitation even though transmitted activity may not take place. As he put it, 'We arrive at the idea of a 'cathected' neuron filled with a certain quantity, though at other times it may be empty' (Freud, 1887–1902, pp.358).

Another innovation of Freud's project related to the passage of electrical excitation from one brain cell to the next. Since this did not always take place, Freud conceived of 'contact barriers' between neurons. Two years after 'the project', Sherrington introduced the term 'synapse' to describe the same structure. Freud's concept of resistance resembles Sherrington's concept of inhibition. Freud wrote 'there are resistances which oppose the discharge of excitation from one neuron to another' (pp.359). It is clearly an aspect of the mechanism of repression.

Freud's next set of speculations concerned whole brain function. He supposed that there may be functionally distinct neuronal systems which are, nevertheless, connected and related. Put another way, he postulated 'inner' and 'outer' neuronal systems. Freud believed that a notional 'stimulus barrier' must operate in order to protect an inner, central, or nuclear system from being overwhelmed by external excitation impinging upon it. His idea of the stimulus barrier was sophisticated. He suggested that not only the amount of excitation, but also the patterning of that excitation determined whether or not the barrier was passed. 'These sense organs operate not only as screens against quantity like every nerve ending apparatus but as sieves, for they only let through stimuli of a particular frequency' (Pribram, 1969, pp.405). The action of the 'sieve' is determined by prior experience, so that as Pribram puts it 'Perceptions result from an interaction between current external stimulation and the residuals of prior experience with similar situations, an interaction modified by concurrent prior and present endogenous excitations' (pp.407). There are times, however, where the 'sieve' cannot work, so that in highly painful or traumatic circumstances excitation slices through the stimulus barrier 'like a stroke of lightning' (Pribram, 1969 pp.407).

Freud's next step concerned emotions and the mechanisms of defence. He saw the nuclear system as containing within it traces of old excitations which will cause certain networks or pathways selectively to activate in the future. Freud supposed that the inner or nuclear system possessed secretory mechanisms so that when 'a certain level of excitation has been reached the neurosecretory cells are discharged' (Pribram, 1969 pp.408). This, in turn, results in the production of more of the chemical substances that stimulate the internal receptors. The organism needs to rid itself of this chemical stimulation. In addition it is required to develop defences against excessive internal excitation and against the release of neurosecretions. These defences depend on rechannelling the excitation. We now return to the hysteria model.

Hysteria depended upon the idea that a memory of an unacceptable sexual event was repressed, so that the neurosecretions were, as it were, dammed up, creating noxious effects. The purpose of psychotherapy was to cause 'resistance to melt away and this enabling the circulation to make its way into a region that has hitherto been cut off' (S.E.II, pp.290-1).

The form of 'rechannelling' was characteristic for each neurosis. Freud remarked that 'An impulse towards the discharge of an unconscious excitation will so far as possible make use of any channel for discharge which may already be in existence'

(S.E.VII, pp.53). In hysteria, in which the symptoms are bodily, the individual is conceived as having a 'capacity for conversion' (S.E.II, pp.50), a kind of vulnerability, later referred to as 'somatic compliance', so that the energy or excitation is converted or 'transformed into something somatic'. (S.E.II, pp.49).

It can be seen that the neuronal theory, which was central to Freud's early theorizing, is relatively simple. It depends upon the notion that the impulse to discharge nervous system excitation was blocked and so, in order to overcome the resultant tension, it was necessarily rechannelled along alternative nervous pathways creating symptoms. In subsequent years Freud built his evolutionary ideas into this basic model.

Freuds early Darwinism was replaced by an adherence to Lamarck. His followers protested against this, but Freud was adamant that the biologists of his era were simply wrong in rejecting Lamarck (Sulloway, 1979). Freud believed that evolution determines not only the physiology, the anatomical characteristics and the behavioral repertoires, but also the history of the species. He believed that we were all developmentally condemned to repeat the history of mankind.

In the early phases of what we choose to call civilization, in the Middle East and in Greece, infanticide was common and so also were the stories of kings who had killed their fathers and married their mothers. Although Freud does not refer to it, an example is provided by a series of Parthian Kings (70BC–AD4). The first two generations were murdered by their sons, who then gained the kingship. In the third generation the king, Phraates IV, was murdered by his wife, who then married her son, after which they ruled together. The Sophoclean drama, Oedipus Rex, a story of this kind, was written nearly half a millennium earlier, and taken from Homer. Freud considered that all males go through a period early in their development in which this phase of the early history of civilized man is replayed. The child wishes to engage sexually with his mother and to kill his rival, the father. The later development of man, Freud conjectured, had led to the emergence of higher constraining forces which no longer allowed these original impulses to be expressed, but which permitted society to hold together. The neuronal theory was now reinterpreted in terms of this larger model. The nervous system excitation involved aggressive and sexual drives. Since they were of an Oedipal kind they were unacceptable and had to be repressed during the individual's later, more mature stages of development. Nevertheless, there had to be a discharge for the internal excitation of libido (i.e. sexual energy) which was the result of these drives. A rechannelling had to occur.

Since all people had to live through this developmental phase, there had to be 'normal' mechanisms for the discharge or rechannelling of libido. An example in waking life included the phenomenon of sublimation. Freud wrote that the sexual instinct 'places extraordinarily large amounts of force at the disposal of civilized activity, and it does this in virtue of its especially marked characteristic of being able to displace its aim without materially diminishing its intensity. This capacity to exchange its originally sexual aim for another one, which is no longer sexual but which is psychically related to the first aim, is called the capacity for sublimation' (S.E. IX, pp.187).

The most important means, however, of the discharge of instinctual drives, so Freud believed, was through dreaming. The dream allowed the excitation aroused by unacceptable sexual wishes to be discharged in fantasy. Through the symbolic mechanisms of 'condensation', in which disparate ideas, images, etc., are fused together, and 'displacement' in which an idea or image, e.g. of a person, is displaced on to another, the wishes were seen as being acted out in disguise, so that the dreamer does not become excessively anxious and is not awoken. In this way, dreams are the guardians of sleep (S.E. XV, pp.143).

Freud revised his original topographical description of brain-mind in the 1920s. His new system — called the 'structural model' — was tri-partite, consisting of ego, id, and superego. Ego was the center of the system.

In essence, ego consisted of consciousness and an elaboration of the 'stimulus barrier'. Freud conceived ego as a neurophysiological system mediating between inner and outer worlds. It was considered to monitor inner and outer events in order to judge their significance, to determine whether they reached consciousness, and to decide upon appropriate responses to stimuli (S.E. XXIII, pp.145–6).

The 'id' replaced the system of the unconscious. It was seen as the repository of instinctual strivings, or drives, which the ego largely kept in check. The judgement about expression, or otherwise, of these drives depended upon the 'superego', a system of prohibitions derived from experience, particularly parental injunctions.

Although the theoretical and practical aspects of Freud's work have come under increasing criticism in the last few years, his opus should be considered relative to his own time. His attempt to build a neural model of mind was truly pioneering. It is only recently that similar attempts have been made. These neural models have the benefit of a considerable amount of neurophysiological data which was not available to Freud. However, his 'ego' was a neurophysiological construct. It would now be possible to create a model of ego, based on his description, which has a sound basis in neurophysiology.

At this point in history, it is not easy to explain the vogue and cultural impact of Freudian theory compared with that of the Salpetriere school. Nevertheless, what was popularly gleaned from his theories must have had a profound resonance with western society at that time.


## OBJECT RELATIONS

The theory of ego became the dominant psychoanalytic orthodoxy during the 1930s. This was particularly so in the United States, where the movement was headed by Heinz Hartmann. Freud's daughter, Anna Freud, maintained the tradition in Britain. Both these analysts focused on the notion of defence. Nevertheless, during this period a number of psychoanalysts found shortcomings in the original theory. Fairbairn, for example, considered that man was not primarily motivated by pleasure or the need to avoid unpleasure. Rather, he believed that the individual was impelled towards others. Put another way, the main drive of a human being as 'subject' was towards another as 'object'. It was now known that disturbances of relationships formed an essential background of those who sought help from psychotherapists. The 'object relations' theorists considered that the central representations of others in these patients were distorted, so leading to expectations of others which disrupted the relationship. These representations can be conceived in terms of traumata which are recorded in memory systems which are more 'primitive' than those underpinning ordinary consciousness (Meares, 1995).

Distortions of self and others are the principal focus of Cognitive Behavior Therapy (CBT), a treatment having a different theoretical basis. This theory is combined with object relations in the therapeutic approach of Anthony Ryle (1990) who has introduced what he terms 'cognitive analytic therapy'. Essentially, distressing target symptoms (attributes) are reattributed by the subject into a more positive context, thereby re-shaping their dysfunctional patterns of behavior (their representations of Self in relation to others).

The addition of object representations to the earlier theory of Freud was clearly important and necessary. Melanie Klein was the most influential amongst the object relation theorists. Her work has now become a dominant theoretical system in the United States due to the influence of Otto Kernberg, who has modified Kleinean language to make it more presentable to a larger audience.

The new theory in essence retained much of what was essential of Freudian theory. However, the unacceptable sexual and aggressive drives were now seen to be malign in a somewhat different way to that suggested by Freud. Klein believed that the sexual and aggressive strivings of the child had damaging implications due to the magical thinking which is characteristic of the child's reasoning before the age of about four. The problem with these strivings was not so much that they were unacceptable to society, but that they could harm not only the real objects who are in the world, but also their inner representations. As a consequence, defences were needed in order to avoid this disaster occurring.

The main theorists in the object relations school were British. They included not only Klein and Fairbairn, but also Michael Balint, Wilfred Bion and D W Winnicott. The work of John Bowlby on attachment theory gave some preliminary scientific basis to conceptions of object relations. Bowlby's approach was steadfastly Darwinian.

Of this group, Winnicott most clearly provided the bridge to a theory of self. His ideas can be seen as parallel to, and in some cases, precursors of, the work of Kohut. He considered that Klein and Freud had underestimated the significance of the environment in the development of the child. He made the famous remark that 'there is no such thing as a baby'. He meant that a baby could not be considered as an isolated one-person organism. A baby depended for the provision of many of its functions upon the social environment.

## SELF

The object relations theorists added an important new dimension to Freudian theory. Memories or representations of others are clearly an important determinant of how the individual functions. However, an essential component of any adequate theory of mind was still missing. Because psychoanalysts had not clearly distinguished between ego and self, they assumed that Freud's references to ego were equivalent to self. When Heinz Hartmann distinguished between these two terms, i.e. between ego and self, in the late 1930s, the way was made clear for the introduction of the latter term. It was one of his proteges, Heinz Kohut, who introduced this term into psychoanalytic discourse. He added a necessary missing element to psychoanalytic theory. He saw development as double, going along two lines, one concerning representations of others and the other concerning self. However, he believed that the fundamental system was that relating to self development. He conceived severe personality disorder as a manifestation of a relative failure of the individual to develop a sense of self.

Kohut elaborated Winnicott's idea that the child's psychological maturation required a facilitating environment provided by caregivers. He considered that the child must feel, at least at times, that the parent is something like an extension of his or her own personal reality. The parent 'fits in' with the child's core experience. This attunement to a child's emotional life leads the parent to be conceived not entirely as an object in relation to the child as subject, but rather, to use Kohut's term, as a 'selfobject'.

The selfobject is the centerpiece of Kohutian theory. However, he found himself unable to define the elusive entity of self, although he could describe its qualities such as cohesiveness and continuity. Furthermore, he declined to distinguish between 'self representation' and a 'sense of self' (Kohut, 1977, pp.311). However, such a distinction is crucial. Self representation must be seen as a more or less stable but potentially mutable aspect of the memory system, dependent on relatively enduring protein charges within neurons. This configuration resembles what Erikson had called 'identity'. On the other hand, the sense of self, which is implicit to Kohut's theory, is shifting, and not necessarily a permanent feature of psychic life. The Kohutian self can attenuate, break up, or even disappear.

Kohut's difficulty in defining self was shared by Jung, who had spent his whole life immersed in the clinical significance of the subject. In the end, he said that his approach had been rather like a circumambulation around unknown factors. If, however, self psychology has any scientific future, it must begin with some definition of what self is. For this reason authorities in the clinical and developmental field have returned to the definitive descriptions of William James. William James, the brother of the novelist Henry James, was intrigued by the phenomena of psychic life and grappled ceaselessly with the problem of trying to capture these phenomena in words. His writing is simple, but the ideas are subtle and profound. James was a considerable philosopher, who, it is said, influenced Wittgenstein. James described a self, which in his parlance, is 'duplex'. It consisted of two poles, 'subject' and 'object'. The subject was the pole of awareness; the object pole was made up of the images, feelings, thoughts, memories, fantasies, etc., which make up psychic life. James called this experience 'the stream of consciousness'. The metaphor, then, of the stream of consciousness is the starting point for attempting to understand the development of self, which Kohut had postulated was deficient in those with severe personality disorder. The eminent developmentalist, John Flavell, in a recent study with his colleagues, showed that the stream of consciousness is not discovered by the child until the age of about five, six or seven. I have suggested that an essential developmental precursor to this experience is what Piaget called the child's symbolic play (Meares, 1993).

Symbolic play does not take up a large part of the child's day. Nevertheless, it appears to be essential for development. During this behavior, the child, who is, say, about three, is absorbed in playing with toys or other objects in much the same way as an adult who is 'lost in thought'. As the child plays he or she chatters. A story, of a limited kind, is usually being told. The language has a curious form which is essentially nonlinear. The famous Russian psychologist Vygotsky described it. It is an abbreviated language, which makes jumps according to associations rather than logic. The links are made by analogy, emotion, and other associative triggers. At times, elements of speech run into each other in the manner of condensation. In extreme cases the language is incomprehensible, suggesting that it is not designed for communication. It seems not unreasonable to suppose that it is a language that is designed to represent, and so to realize, a sense of an inner life which might be called self. This language differs qualitatively from another language which the child usually engages in. This language is for dealing with the environment. It is triggered by alerting stimuli and is externally oriented. It is logical, linear and clearly adaptive. It has a communicative purpose. The child asks for information, reassurance, nurture, etc.

Before the age of three the child oscillates, often very quickly, between two modes of experience which are governed by different forms of mental activity manifest in different kinds of language. One is nonlinear underpinning symbolic play; the other is linear governing adaptive communicative behavior. Each form of mental activity, it must be presumed, is underpinned by a specific 'module' of brain function.

At about the age of five, six or seven, the curious speech accompanying symbolic play disappears and, so Vygotsky believed, becomes internalized to become the language of an inner life. It is consistent with the Jamesian metaphor of the stream that this language should be nonlinear.

In adult conversation, the kind of language used by the child engaged in symbolic play is not found in pure form. However, the mental activity which underpins this language is likely to be reflected in such social and cultural productions as myth and folklore. Jung made these cultural phenomena central to his explorations of self. These narratives, although much more complex, are not unlike those told by the child during symbolic play. Jung's studies lead him to consider that they have a characteristic dynamic form, which he called archetypal. Jung believed that the structuring of these narratives was a reflection of the brain's manner of organizing the sensory data of everyday life. Since human lives resemble each other, certain archetypal themes are common to all cultures. The central archetype is the self and its principal theme is the myth of the hero.

Ordinary adult conversation, as previously remarked, does not reflect in pure form the mental activity underpinning the narratives of the archetypes and of symbolic play. Nevertheless, most conversations involve a combination of two kinds of languages which are the markers of two different kinds of mental activity. The linear form of language which is clearly designed for communication, adaption, and coping with the social environment, is dominant. Mingled with it are elements of the second, non-linear mental activity characteristic of inner life, or self. As a consequence it is usually possible 'to discern within social speech and the form of language built for discourse with the outer world, the embedded elements of another kind of speech through which self can be visualized' (Meares, 1993, pp.183). However, in those whose development has been severely disrupted, e.g. the borderline personality, the mental activity to do with innerness can barely be demonstrated, at least at first. The individual is as if addicted to stimuli, not only of the world, but coming from his or her own body. The content of the talk is about the problems of their life events, symptoms. The form of the language which is remorselessly linear.

Therapy is directed towards establishing or re-establishing the self system.

Therapeutic improvement is indicated by a change in the structure of the therapeutic conversation, which becomes more complex. In simplest terms, it could be said that the aim of therapy is to move the therapeutic conversation from a state of low complexity to a relatively higher state of complexity.

The above ideas suggest that words are the markers of self, so allowing it to be studied. When the mental activity which underpins the Jamesian stream of consciousness comes into operation it will be manifest in a particular form of language. One approach to the study of self, at our present stage of development, might be undertaken by a combination of linguistics and the new mathematics which relates to complexity and non-linearity. This possibility has been anticipated by a flurry of papers in the last few years which have attempted to use the metaphors of chaos theory and quantum theory as a means of conceiving mental processes. Indeed, such metaphors were invoked by Jung in his attempt to describe the complexity of self. At least one of his formulations resembles the modern notion of the 'strange attractor' (see chapter 10a for details of attractors).

Another relevant concept to self provided by metaphors from the new mathematics, may be the fractal theory of Mandelbrot. This kind of complexity arises through the process of 'iteration', which, when translated into the neurophysiological setting, depends upon the evolution of feedback loops from higher centers to lower levels in the central nervous system. It is supposed that these feedback systems will be present in human beings but not in other primates. It is also supposed that such a system is seminal to the reflective capacity which is necessary to the emergence of the experience of the stream of consciousness. Mandelbrot's theorem predicts the form of stream-like meandering (Stolum, 1996).

## CONCLUSION

In order to establish the discipline as a science, it is necessary to build a theory of psychotherapy which is based upon falsifiable postulates. The ego psychology of Freud is not readily amenable to testing because it depends upon the concept of the unconscious. The newer theory of self can be studied when it is conceived not as a thing, but as one of the manifestations of a particular form of brain function, the activity of which is reflected in certain kinds of linguistic behavior.

## REFERENCES AND SUGGESTED FURTHER READING

Freud S (1887-1902) *The Origins of Psychoanalysis*: Letters to Wilhelm Fliess, Drafts and Notes: 1887-1902. London, Imago, 1954.

Freud S (1953-73) *The Standard Edition of Complete Psychological Works* (24 vols) S.E. London, Hogarth.

Flavell J, Green F and Flavell E (1993) Children's Understanding of the Stream of Consciousness. *Child Development*, **64**, 387–396.

James W (1892) *Psychology*: Briefer Course. London, MacMillan. Paperback. London, Collier-MacMillan, 1962.

Kohut H (1977) *The Restoration of the Self*. New York: International Universities Press.

Meares R (1993) *The Metaphor of Play*: Disruption and Restoration in the Borderline Experience. Northvale N. J: Jason Aronson

Meares R (1995) Epiodic Memory, Trauma and the Narrative of Self. *Contemporary Psychoanalysis*, **31**, 541–556.

Pribram K. (1969) The Foundation of Psychoanalytic Theory: Freud's neuropsychological model. In:ed. K. Pribram 'Adaptation: *Brain and Behaviour,* **4**, 395–432.

Ryle A (1990) *Cognitive Analytic Therapy: Active Participation in Change.* Chichester, Wiley and Sons.

Stolum H. (1996) River Meandering: A Self Organisation Process. *Science,* **271**,1710–13.

Sulloway F (1979) Freud: *Biologist of the Mind.* New York: Basic Books.

# Chapter 15.

# HUMAN BRAIN IMAGING TECHNOLOGIES

Evian Gordon[1], Chris Rennie[2], Arthur Toga[3] and John Mazziotta[4]

[1] The Brain Dynamics Centre, Westmead Hospital Department of Psychological Medicine,
The University of Sydney
[2] Department of Medical Physics, Westmead Hospital School of Physics, The University of Sydney
[3] Laboratory of Neuro-Imaging, Department of Neurology, UCLA School of Medicine,
The International Consortium for Brain Mapping
[4] Brain Mapping Division, Department of Neurology, UCLA School of Medicine,
The International Consortium for Brain Mapping

Until the 1970s, examining the human brain as a system was mainly undertaken post-mortem in clinical studies of patients with brain lesions or stimulating the brain during operative procedures. With advances in physics and computing, a range of computerized brain imaging technologies have been developed. Innovations in physics made accessible wave lengths, such as X-rays, radiowaves and gamma rays, which coupled with advances in computing technology allowed acquisition of the volume of data necessary to generate images of brain structure and function.

## TESTING HUMAN BRAIN MODELS

Since Broca's description of a localized speech area in the brain last century, the predominant view has been that higher brain function can be understood as hierarchical assemblies of 'centers' with consistent localized anatomy. This model remains entrenched in clinical practice, where it is used to explain relationships between brain lesions and symptomatology.

A number of the chapters in this book show that whilst there is considerable evidence from clinical, pathological and physiological research implicating a relationship between brain region deficits and clinical symptoms for specialized sensory-motor-speech processes, there is no one-to-one relationship between anatomical site-brain function-behavior for higher order processes (particularly in the association cortex). The theoretical status and clinical utility of center based localizationist models is therefore being called into question.

Current models suggest that brain function may be better characterized as consisting of ongoing interactions across scale, between localized and distributed neuronal networks, which are engaged according to situation and task demands. Behavior being not the result of activity at a center level, but adaptive tempero-spatial processing of multiple neuronal networks operating in parallel. Behavior would therefore be associated with multiple neural networks, rather than specific anatomical sites. In The Big Picture in the book, it is suggested that dichotomous conceptualizations of the brain, such as specialized versus holistic models, are misleading, since both readily co-exist.

Other chapters in this book exemplify the emerging models across disciplines, and scales of function that make explicit predictions about relationships between brain function and information processing. Many models that address the whole brain as a system are now able to be tested using brain imaging technologies. The 'goodness of fit' between these whole brain *theoretical* models and brain imaging *measures* will increase or decrease the likelihood that the model is true or false. Measurement is after all the final arbiter of the validity of any model. The first realistic numerical models now also show preliminary promise to shed light on measures of whole brain function in a fundamental manner, including assessment of possible mechanisms of action.

## IMAGING BRAIN STRUCTURE AND FUNCTION

A description of each brain imaging technology is provided, followed by a discussion of their current status and emerging future directions.

### A Description of Brain Imaging Technologies

A brief description of each technology is provided. Figure 15.1 provides a summary of the spatial (where) and temporal (when) dimensions that these technologies are able to explore in the brain. They are subdivided to make explicit their complementarity.

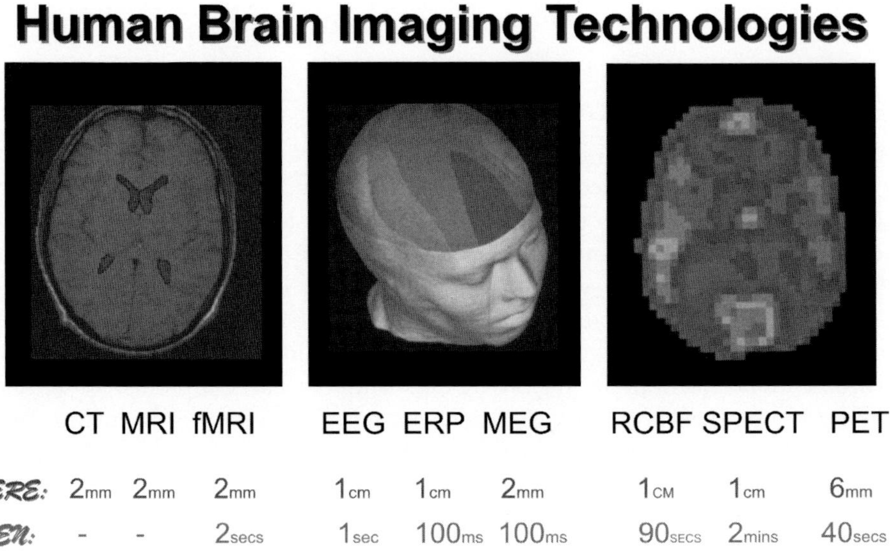

# Human Brain Imaging Technologies

| | CT | MRI | fMRI | EEG | ERP | MEG | RCBF | SPECT | PET |
|---|---|---|---|---|---|---|---|---|---|
| *WHERE:* | 2mm | 2mm | 2mm | 1cm | 1cm | 2mm | 1CM | 1cm | 6mm |
| *WHEN:* | - | - | 2secs | 1sec | 100ms | 100ms | 90secs | 2mins | 40secs |

**Figure 15.1.** Spatial and temporal dimensions of brain imaging technologies.

# Structural Brain Imaging

## Computed Tomography (CT)

CT relies on the fact that X-rays passing through the brain lose energy in proportion to the density of tissue through which they travel. CT scanners generate a fanbeam of X-rays that pass onto an arc of detectors that swing around the head. The attenuation along each X-ray is detected and the density of the brain structure is reconstructed from each two dimensional slice scanned from the fanbeam. Limitations of CT include the contrast resolution between grey and white brain matter, which is poor. The partial encasement of the temporal lobe by skull also smears the X-rays and reduces visualization of brain in this region. Bone itself is clearly visualized. The spatial resolution of transverse slices of brain structure is 1–2mm (but is usually 5mm in practice since the amount of ionizing radiation that a person is exposed to is proportional to the number of transverse brain slices).

## Magnetic Resonance Imaging (MRI)

When the head is placed within a static magnetic field (several thousand times stronger than the earth's magnetic field), naturally occurring elements in the brain such as hydrogen (which constitutes 60% of the atoms in the brain) act like bar magnets and align themselves in this magnetic field. A brief pulse of radiofrequency waves is then passed through coils positioned around the head, resonating the hydrogen atoms in the short-wave radiofrequency range. The time taken by the hydrogen atoms to realign themselves into the static field is detected by radiofrequency receiver coils around the head, and is faster in white than grey matter. This differential realignment time provides the basis for the excellent contrast resolution between grey and white matter obtained with MRI. Coronal, transverse and sagittal slices can be derived from the three dimensional volume of brain structure obtained by systematically changing the orientation of the static field. The spatial resolution is 1–2mm (routinely 5mm). Contrast resolution of specific brain structures can be further enhanced by modifying the imaging parameters (pulse sequence and delays). Additional advantages of MRI over CT include no ionizing radiation and no bone artifact. Atoms other than hydrogen (such as phosphorous and sodium) may also be imaged (and form the basis for Magnetic Resonance Spectroscopy or MRS), but their prevalence in the brain is far less than hydrogen and consequently their spatial resolution is poor.

## Functional Magnetic Resonance Imaging (fMRI)

fMRI allows superimposition of high spatial resolution images of brain structure (generated via conventional MRI) with high temporal resolution images of cortical blood flow in response to activation.

The superimposed images of blood flow are measured as the extent of change in deoxyhemoglobin to hemoglobin in specific brain areas, because deoxyhemoglobin is paramagnetic and therefore generates local magnetic field distortions in the magnetic resonance signal. Following stimulus activation to a subject, oxygenated blood flow to required areas produces a change in the concentration gradient of oxyhemoglobin relative to deoxyhemoglobin, and a local MR signal.

The time resolution of these changes is approximately 2 seconds in primary sensory cortex (and a signal change of approximately 10% with activation), and slightly longer in association cortex (and smaller signal changes of about 0.5–2%). Recent advances in fMRI show promise to reflect aspects of activation over shorter time scales in response to discrete stimuli (single-trial fMRI).

## Functional Brain Imaging

### Electroencephalography (EEG)

The EEG primarily arises from the summation of potentials in thousands of synchronously active dendrites in cortical neurons (particularly pyramidal cells which are aligned in columns perpendicular to the cortical surface and their summated activity is thereby discernible). EEG currents are measured non-invasively using recording discs on the scalp (relative to a reference recording on the head) and reflect synchronized and desynchronized oscillations of the overall cortical electrical activity (and subcortical modulations) in the brain. The time resolution is in the order of a second.

A small number of fundamental EEG rhythms emerge (0.5–3.5 Hz Delta; 4–7.5 Hz Theta; 8–12 Hz Alpha; 12+ Hz Beta; 40 Hz Gamma). The EEG exhibits transient steady states that are perturbed by incoming stimuli or subcortical modulations, at which time they rapidly switch to a new transient steady state. The tempero-spatial EEG distributions of fundamental rhythms and their power change with physiological state and processing tasks (for example, Delta with slow wave sleep, Theta with learning, Alpha with idling, Beta and Gamma with brain activation).

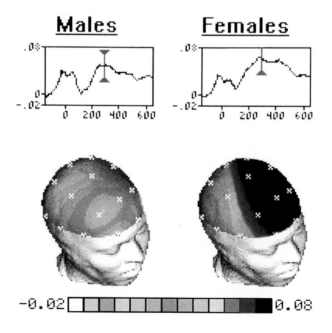

**Figure 15.2.** Evoked Gamma (40 Hz) activity in a group of 50 male and 50 female subjects. Gamma is thought to reflect integration of disparate neural network activity. Females have been shown to have relatively greater interconnected brain networks than males, reflected here in black, as enhanced Gamma activity. *Courtesy of The Brain Dynamics Centre, Westmead Hospital.*

## Event Related Potentials (ERP)

To measure ERPs you add to the EEG setup a computer controlled stimulus. ERPs are transient electrical potentials occurring on a time scale of milliseconds, time-locked in response to discrete sensory, motor or cognitive stimuli. The ERP is however much smaller than the ongoing EEG in which it is embedded. Traditionally, multiple stimuli are presented and the individual time-locked responses are averaged to enhance the ERP. Each peak or trough of the ERP is known as a component. Exogenous or early ERP components occur within the first 100ms after the stimulus and are routinely used in the clinical setting to reflect the integrity of sensory neural pathways. Endogenous or longer latency ERP components are associated with preparatory (CNV), attentional (N100), orienting (P300a), echoic memory (MMN), decision making (P200), arousal and response execution (N200), context (P300b), incongruency (N400) of the stimuli and integration of information processing (evoked 40 Hz or Gamma activity).

ERPS are therefore well placed to test models of cognition. A fundamental common dimension between ERPs and psychological constructs is that of time–the concomitant variation of psychological and neurophysiological processes. ERPs provide the highest temporal resolution of brain imaging technologies, and are therefore used as real-time markers for both psychological and physiological events. Most cognitive processes occur, after all, on a time scale of tens of milliseconds, rather than seconds.

## Magnetoencephalography (MEG)

Electrical currents in the brain are rapidly volume conducted from their source and then smeared by the different electrical resistance's inherent to the coverings of the brain, the skull and the scalp. This smearing, as well as the fact that there is no unique solution to three dimensional source localization from two dimensional scalp measures, makes it difficult to determine with certainty from scalp recordings the sources of electrical brain function.

MEG makes use of the fact that a magnetic field surrounds all electrical currents. Unlike electrical currents, however, magnetic fields are not smeared by biological tissue nor are they volume conducted. Rather, they fall off rapidly and predictably from their origin. They therefore emerge from the skull close to their source and relatively undistorted. These magnetic fields can be detected using a superconducting quantum interference device (SQUID) coupled to a sensor coil (gradiometer) placed just above the scalp. The distribution of the magnetic fields measured outside the head can be used to accurately determine their electrical source within the brain. An additional dimension of MEG is that electrical currents are measured relative to a reference, whereas magnetic fields require no reference and therefore provide absolute rather than relative measures of electrical brain function (this is an important advantage when comparing measures across subjects). MEG is therefore complementary to EEG/ERP for localizing electrical activity in the brain.

# Radioisotope Measures

## Regional Cerebral Blood Flow (RCBF)

RCBF is measured using gamma emitting substances, usually Xenon-133, a chemically inert gas that is introduced into the body via inhalation or intravenous injection. The gas saturates the brain and over the ensuing 11 minutes is cleared from the brain at a rate proportional to blood flow (which is highly correlated with brain metabolism). The grey matter of the cerebral cortex is well perfused and fast clearing, particularly in the 30–90 seconds after injection. It is this cortical activity (rather than subcortical less well perfused and slow clearing white matter) that this method primarily reflects. The gamma ray counts reflecting activity and clearance are measured using stationary gamma-scintillation detectors (up to 254) positioned around the head. Topographical images of cortical RCBF are derived from these measures. Temporal resolution is commonly determined as the 30–90 seconds after injection. Spatial resolution in a 254 detector system is approximately 1cm of cortex.

### Single Photon Emission Computed Tomography (SPECT)

The radioisotopes used in SPECT are not normally metabolized in the brain (for example Iodine, Technetium and Thallium). These substances are labeled (according to a brain function of interest such as blood flow or the distribution of a specific neurotransmitter) and injected intravenously. The labeled radioisotope is trapped within 2–3 minutes in or onto the neurons, in proportion to the activity of the labeled substance. The trapping of the isotope lasts for sufficient time to allow the rotating gamma camera(s) to acquire the counts of radioisotope distribution throughout the brain. From this three dimensional information, transverse, coronal and sagittal reconstructions of cortical and subcortical activity are derived.

Temporal resolution is approximately 2 minutes. Spatial resolution is 6mm with triple headed cameras. Semi-quantitative analysis, such as brain region of interest to cerebellum ratios are undertaken.

### Positron Emission Tomography (PET)

A positron-emitting isotope such as $F1^{18}$ (half-life 110 minutes), $C^{11}$ (half-life 20 minutes), $N^{13}$ (half-life 10 minutes) or $O^{15}$ (half-life 2 minutes) is labeled to a biochemical substance that is naturally used by the brain (such as a neurotransmitter, glucose, water, an amino acid or a drug) since a positron is antimatter (an antielectron) and once injected into the body, positrons collide with electrons leading to the emission of a pair of gamma rays which are given off in opposite directions. The points of coincidental emission in the brain are detected by rings of gamma detectors positioned around the head. This information is used to construct images reflecting slices of cortical and subcortical distribution of the brain's activity.

The temporal resolution is in the order of minutes (40 seconds for oxygen, minutes for other radioisotopes). The spatial resolution is 4–6mm depending on the system. PET measures of brain function can be quantitative and analysis requires measurement of changes in venous and arterial concentrations of the radioisotope over the course of the study.

The relatively short half-lives of positrons compared to radioisotopes used in RCBF and SPECT, require an on site cyclotron to produce the positrons, escalating both the physical complexities and cost of this tool of brain exploration. Over the past decade SPECT has encroached into domains of brain exploration previously considered exclusive to PET, such as assessing a range of neurotransmitter and drug distributions. PET remains the pre-eminent quantitative technology in exploring in-vivo brain chemistry and drug receptor activity. However, advances in PET are likely to be carried into the clinical setting with SPECT.

# CURRENT STATUS OF BRAIN IMAGING TECHNOLOGIES
## Complementarity

While each technology is sophisticated and has a particular strength, the measures obtained are still limited compared to the complexities of overall brain function. Considered together however, they provide complementary indices of brain structure and function. Each brain imaging technology has a different strength: examining either brain structure, electrical activity or chemical activity. In addition to complementary measures of brain structure and function, these technologies provide information at different time scales (of *when* brain activity occurs), as well as different spatial resolution (of *where* this activity occurs).

CT and MRI provide detailed scans of brain structure that are relatively unambiguous to interpret. These scans however, provide limited inference of function, since there is not a one to one relationship between brain structure and function. In contradistinction, changes in brain function that do not cause structural damage may result in behavioral changes. Furthermore, the integrative functions of the brain, stages of information processing, brain chemistry and drug activity, all can only be assessed by the examination of brain function.

EEG, ERP and MEG allow examination of electrical brain function with very high fraction of a second resolution in time (but poor spatial resolution). The nuclear medicine measures RCBF, SPECT and PET reflect brain metabolism and chemistry with somewhat better spatial resolution, particularly of subcortical activity in the case of SPECT and PET (but poor temporal resolution). fMRI brings together a measure of blood flow with high temporal as well as spatial resolution (but not with the temporal resolution of electrical measures and does not measure chemistry).

There are many specific examples for complementary use of these technologies. CT provides clear images of bone and recent hemorrhages, whereas MRI provides better detail of grey and white brain matter necessary for detection of small tumors and subtle disturbances of brain structure.

Magnetic field measures outside the head detect only that portion of electrical brain function that is orientated parallel to the skull surface (in cortical sulci). Electrical measures, on the other hand, primarily index activity in sources orientated radially with respect to the scalp (in cortical gyri). Therefore, simultaneous assessment of ERP and MEG will provide complementary spatial information of networks underlying time-locked brain function.

The SPECT and PET radio-isotope based imaging techniques allow measurement of metabolic activity. In order to study normal brain-behavior relationships, activity is measured during specific activation tasks. By mathematically computing the differences in the scans between activation conditions, changes in brain responses are correlated with changes in the task conditions. Co-registration of these changes in brain function with the same subject's MRI and a second co-registration onto a standard digital brain atlas (to normalize the data) allows identification of the brain regions which are primarily activated.

In summary, fMRI is increasingly being employed in cognitive activation studies, with PET and SPECT better suited to assess brain chemistry and drug receptor activity, and electrical measures probing increasingly refined temporal aspects of cognition.

## Clinical Applications

The most commonly employed brain structural technologies in the clinical setting are CT and MRI, which are used to exclude a biological illness (such as a tumor or stroke) and explore the relationship between brain structure-function-behavior.

The primary clinical use of EEG is to help diagnose epilepsy. It is also used to examine sleep disturbances and quantified EEG shows some promise as a means of non-invasively evaluating clinical subtypes and the effects of medication on brain function.

Early ERP components reflect the integrity of the sensory pathways and are used routinely to diagnose disorders such as multiple sclerosis, and to identify abnormalities in the sensory pathways. Late component ERPs have been used as aids to diagnose neuropsychiatric disorders, sub-types of disorders and to evaluate the effects of medication on aspects of cognition. But simplistic analysis techniques, small control databases and poorly evolved models concerning the nature of late components ERPs have so far limited their utility in the clinical setting.

PET and SPECT studies have shown that specific activation tasks are associated with preferential activity of certain loci within brain networks. Clinical studies have shown consistent patterns of changes in brain function associated with different disorders. For example, decreased activity in the temporal and parietal lobes in Alzheimers dementia, decreased frontal activity in schizophrenia and percentage uptake in different neurochemical receptor activity associated with changes in clinical symptoms in various disorders (see summary in figure 1.17 in The Big Picture).

There is also often a number of brain imaging disturbances found in the same disorder. Broadbent (1958) suggested that early stages of processing lead to downstream disturbances of function, and this possibility needs to be examined in individual patients to distinguish cause from effect. The importance of exploring such brain dynamics is also highlighted by the 'kindling model', that with successive clinical episodes of a disorder there may be large non-linear changes in stability of brain function and the effect of medication (Post 1992). Considering different possible patterns of adaptive brain function over time should be an important dimension in interpreting brain imaging measures.

Since changes in brain function antedate symptomatology, the use of these technologies will increasingly be used prospectively to identify pathology, clinically to identify sub-types and practically to assess the effects of medication on the brain as a system. However, fundamental progress in these applications awaits better models that reflect a clearer understanding among brain structure-function-symptomatology interrelationships.

## Future Directions

There are a number of future directions in the burgeoning field of brain imaging, a selected few are considered below.

1) Multimodality. There are now increasing efforts to bring together in the same individual appropriate combinations of brain imaging technologies, to exploit their complementarity. There are many technical refinements in image registration and multimodal data analysis that are under development. What is still being resolved is the cost-benefit of which combination of technologies are most appropriate for different clinical and basic research needs.

2) Activation studies. Appropriately designed activation tasks are able to preferentially activate brain networks underlying aspects of cognition. But standardized neuropsychological tests cannot simply be grafted into the brain imaging arena — they need to be carefully modified to be consistent with the physiological time-scales of measurement provided by each brain imaging technology.

We should also be circumspect about some subtraction studies (for example, when a 'baseline' imaging study is subtracted from an 'activation' study to expose networks engaged specifically with the activation). Even in the simplest tasks, adaptive brain function may change significantly in multiple ways on two consecutive recording sessions. Moreover, in clinical studies when disturbances are reported in some specific 'high level' function (such as memory, abstraction, planning) it is prudent to ensure that an underlying 'lower level' dysfunction has been precluded.

Within-subject studies, in which systematic levels of activation are undertaken in the same subject who acts as his own control, offer an additional useful approach to activation and have the potential to avoid many of the confounding variabilities of between-subject studies.

3) Control for performance. Many studies show between-group differences in brain function that are confounded by performance. The clinical goal is to find distinctive patterns of fundamental neuronal network dysfunction in disorders that are not simply a consequence of differential motivation and performance by the subject. The same could be said for differential arousal, habituation and attention. You do not need a brain imaging technology to show abnormalities in performance, arousal, habituation and attention *per se* — since reaction time, electrodermal and eye movement measures could in many cases be quite adequate to answer such questions, and be more cost effective.

4) Simultaneous examination of central and autonomic function. Alternatively, such within-trial physiological and behavioral processes (arousal, attention, habituation and performance) can be interpreted with more confidence with respect to brain imaging measures if they are acquired simultaneously with measures of brain function (because brain function is so dynamic, assumptions of stability over time are fallacious). The simultaneous measurement of these variables with brain function can readily occur, for example, measuring arousal and habituation via electrodermal or heart rate activity, monitoring attention by tracking eye movement and measuring performance with reaction time.

5) New approaches to analysis of the data. Most functional studies examine *averaged activity* across the trial, which is a solid and appropriate first step. But systematic processes (such as orienting, learning, 'Automatic' or 'Controlled' processing) may vary in a systematic manner across the trial, and can be separately sub-averaged and assessed (providing complimentary information to the average measures). The overall patterns of change in brain function can also be examined (see chapters 10 a and b). In addition, our limited understanding of brain function may not have as much to do with what we have measured, as with the level of sophistication with which we have analyzed these complex signals. Multidisciplinary involvement in human brain science is resulting in new mathematical, signal processing and statistical approaches, that extract more information from these complex measures.

6) Specificity of findings, sub-types of clinical disorders, assessment of symptom profiles associated with findings and longitudinal examination of patients before and after treatment, all will help to reveal more distinctive patterns of brain dysfunction in neurological and psychiatric disorders.

7) Databases. There will ultimately be a family of brain imaging databases world-wide. We present one example, the International Consortium for Brain Mapping (ICBM).

## The International Consortium for Brain Mapping (ICBM)

Neuroscience data is collected at an ever accelerating rate and volume. Sub-specialties in neuroscience are the rule rather than the exception and there are approximately two hundred specialty journals in neuroscience. It is an overwhelming amount of information. A lot of the information is lost, we just cannot keep track of it. We can only publish a certain small fraction of what we collect. In addition, an investigator might have the capacity to collect very meticulously gathered data about a specific variable but without a context in which to place it and a reference system to use for correlating their data with that of others. However, if we had standardized ways to store, integrate and register data, we could examine information across laboratories, as well as see this data in a broader context.

Efforts are therefore underway in programs such as the ICBM (which is part of The Human Brain Project — see Koslow and Huerta, 1997) to develop a suite of electronic atlas's of brain structure upon which all multimodal measures of brain function, as well as brain chemistry and effects of drug activity, can be co-registered.

One goal of the ICBM is to build a probability analysis of the human brain to which other researchers can interface their multimodal data, thereby increasing the confidence limits of the data, and the scope of data that is available for examining interrelationships and for testing models of the brain.

This project emerged from frustration with current atlases of the brain made up from one individual, which have varying degrees of accuracy. For specific procedures they may be effective, but if we really want to learn about the brain and organize data that comes from many individuals, then we need to know about the variance between individuals. The basic assumption is that no two brains are really the same, both structurally and functionally. The ICBM are building a population atlas where information is expressed as a probability. They have created a co-ordinate system, initially structural, with MRI data from hundreds of normal people. More detailed structural information is also coming from very high resolution anatomical specimens, where the entire head of the human subject is sectioned with hundred micron resolution in all directions. Several stereotactic atlases have been produced for the human brain (for example, Talairach and Tournoux, 1988 in Toga and Mazziotta, 1996) to establish standard spatial references and apply three dimensional co-ordinates for specific structures. The number of sections published in these current atlases is not sufficient for reflecting neuroanatomical complexity. For these reasons the ICBM project has included (in addition to MRI) three dimensional digital modeling based on high resolution image data captured from cryosectioned anatomic specimens.

High resolution anatomic data sets will be able to serve as a frame of reference for the accurate interpretation of normal and clinical data from CT, MRI, fMRI, RCBF, SPECT and PET modalities, as well as the mapping of transmitter substances and other regional biological functional characteristics. A significant challenge will be relating and modeling these metabolic measures with respect to the dynamical measures of electrical activity provided by EEG, ERP and MEG.

Such databases are still in their infancy, with significant issues such as quality control and consistency of activation paradigms still to be resolved. However, coupled with the emerging field of 'Neuroinformatics' (Koslow and Huerta, 1997), which shows the potential to bring together diverse information about the brain into databases (including access via the internet), integration of data (including from brain imaging) for testing of models is becoming available and accessible on an unprecedented scale.

A crucial factor in the appropriate interpretation of brain imaging technology data, is to be cognizant of the limitations of reductionism, particularly in the face of the complexities of the brain as an adaptive dynamical system. As we have re-iterated, there is no one-to-one relationship between anatomical site-brain function-behavior. Much brain imaging data is presented in localizationist and reductionist terms. Beware of brain scientists bearing localizationist ('hot-spots') computerized phrenology color images of the brain. Skepticism should prevail, until the interpretation of such seductive images are couched within the context of the future directions (listed above). Interpretation of all brain images should also be coupled with an understanding of anatomical connectivity, specific mechanisms of brain electro-chemistry and overall organization of brain function — which goes way beyond the platitudes that multiple regions in the brain are connected and interact.

This is not to detract from the fact that a localizationist model has been very useful in specific circumstances for studying sensory, motor and speech networks in the brain. But what needs to be remembered when examining whole brain imaging measures, is that specialized sensory-motor networks constitute only a small percentage of the human brain. Whilst we do not know the mechanisms of mind that integrate the specialized and modular networks of brain function, it is known that of the total hominid brain size that has trebled over the past six million years, the vast majority of this increase has been in the association cortex (see chapter on evolution). In addition, a key feature of the brain is its highly interconnected nature (see anatomy chapter). To understand measures taken from such an associative interconnected system — surely demands a holistic and model driven approach.

See The Human Brain Project on http://www.nih.gov/grants/guide/pa-files/PAR-99-138.html

## CONCLUSION

This is a new era. Some clinical utility has already been achieved with the use of brain imaging technologies. However, to fully exploit the potential of this emerging field requires a more fundamental understanding of human brain macrophysiology.

This book outlines the brain's extraordinary complexity from the molecular to system level. Different properties emerge at different scales of function. Brain imaging technologies reveal aspects of whole brain structure, large scale electrical activity and metabolism. But for all their power and diversity, brain imaging technologies can't image all elements of cognition.

The final crucial step in interpreting brain imaging measures is still the requirement for something in addition to imaging to make the link between the measures and what they realistically might mean. A simple analogy may be the weather. Elements of basic weather patterns such as cloud cover, periodicity and duration of storms can be obtained from satellite photographs and measures on the ground. But they do not reveal the causal mechanisms across scale that underlie these storms which have been derived from theoretical and experimental physics, and testing explicitly possible interactive mechanisms in models. Similarly in the brain, possible mechanisms of function need to be explicated in models.

As attested by most of the chapters in this book, a step-by-step combination of linear and (where appropriate) non-linear approaches will be a fulcrum in the interactive exploration of the goodness of fit between new *theoretical models* and *empirical measures* of human brain function and cognition. The likelihood of increasingly robust outcomes is high, due to the confluence of multidisciplinary activity in model development and testing, that is occurring in the field of brain imaging technologies.

## REFERENCES AND SUGGESTED FURTHER READING

Toga, A. W, Mazziotta J. C (1996) [eds.]. *Brain Mapping:* The Methods. Academic Press, San Diego.

Broadbent, D. E, (1958) *Perception and Communication.* London: Pergamon Press.

Koslow, S. H and Huerta, M. F (1997) *Neuroinformatics: An overview of the Human Brain Project.* Lawrence Erlbaum Associates, New Jersey.

Pfurtscheler, G and Lopes da Silva (1988) *Functional Brain Imaging.* Han Hubers Publishers.

Post, R. M, Weiss, S. R. B, Ketter, T. A, Denicoff, K. D, George, M. S, Frye, M. A, Smith, M. A and Leverich, G. S (1998) *The Kindling Model: Implications for the Etiology and Treatment of Mood Disorders.*

Wade, J, Knezevic, S, Maximilian, V. A, Nubrin, Z, and Prohovnik, I (1987) *Impact of Functional Imaging in Neurology and Psychiatry.* John Libby, London.

# INDEX

(Information in tables and figures has not been indexed.)